Farmland
PRESERVATION

Farmland
PRESERVATION
Land for Future Generations

WAYNE CALDWELL, STEW HILTS,
AND BRONWYNNE WILTON

SECOND EDITION

UNIVERSITY OF MANITOBA PRESS

© The Authors 2017

21 20 19 18 17   1 2 3 4 5

All rights reserved. No part of this publication may be reproduced or transmitted in any form or by any means, or stored in a database and retrieval system in Canada, without the prior written permission of the publisher, or, in the case of photocopying or any other reprographic copying, a licence from Access Copyright, www.accesscopyright.ca, 1-800-893-5777.

University of Manitoba Press
Winnipeg, Manitoba, Canada
Treaty 1 Territory
uofmpress.ca

Cataloguing data available from Library and Archives Canada
ISBN 978-0-88755-820-7 (PAPER)
ISBN 978-0-88755-520-6 (PDF)
ISBN 978-0-88755-518-3 (EPUB)

Cover design by Frank Reimer
Interior design by Jess Koroscil
Cover photo by Paul Kraehling

Printed in Canada

The University of Manitoba Press acknowledges the financial support for its publication program provided by the Government of Canada through the Canada Book Fund, the Canada Council for the Arts, the Manitoba Department of Sport, Culture, and Heritage, the Manitoba Arts Council, and the Manitoba Book Publishing Tax Credit.

Funded by the Government of Canada | Canadä

CONTENTS

PREFACE TO THE SECOND EDITION - VII

INTRODUCTION
Farmland Preservation Perspectives
Bronwynne Wilton - 1

CHAPTER 1
Canadian Farmland: A Fluctuating Commodity
Michael Troughton - 13

CHAPTER 2
Agricultural Land Protection in Quebec
Christopher Bryant, Claude Marois, Denis Granjon, and Ghalia Chahine - 28

CHAPTER 3
Farmland Preservation in Ontario
Wayne Caldwell, Stew Hilts, and Bronwynne Wilton - 46

CHAPTER 4
The Farmland Preservation Program in British Columbia
Barry E. Smith - 64

CHAPTER 5
Learning about the Agricultural and Food System in Your Municipality
J.C. (Jim) Hiley - 94

CHAPTER 6
Smart Growth in Ontario: Getting Ahead of the Future
Gary Davidson - 114

CHAPTER 7
Farmland Protection and Livable Communities in British Columbia
Kevin McNaney and Kelsey Lang - 124

CHAPTER 8
Rural Non-Farm Development and the Agricultural Industry in Ontario
Wayne Caldwell, Arthur Churchyard, and Claire Dodds - 134

CHAPTER 9
Preserving and Promoting Agricultural Activities in the Peri-Urban Space
Nicolas Brunet - 152

CHAPTER 10
Ontario Farmland Trust: Bringing Permanence to Farmland Protection
Matt Setzkorn - 171

CHAPTER 11
Farmland Preservation Policies in the United States
Tom Daniels - 184

CHAPTER 12
Planning for the Future of Agriculture
Bob Wagner - 200

CHAPTER 13
Farmland Preservation in Australia
Trevor Budge and Andrew Butt - 210

CONCLUSION
Farmland Preservation: Land for Future Generations
Wayne Caldwell, Bronwynne Wilton, and Kate Procter - 236

CONTRIBUTORS - 241

Preface to the Second Edition

WAYNE CALDWELL, STEW HILTS, AND BRONWYNNE WILTON

Farmland Preservation: Land for Future Generations speaks to farmland as an essential resource, meeting not only a basic human need but also important fuel, fibre, energy, and sociocultural needs. As land is lost to urban sprawl and other non-farming activities, our ability to produce food and protect agri-environmental resources becomes limited. Farmland preservation is about protecting the agricultural land base for future generations. This need for protection is driven by uncertainty caused by climate change, population growth, food security, energy availability, and a host of other local and global factors. This uncertainty means that there is an ever-growing responsibility to ensure that the actions of today do not compromise the needs of future generations.

This responsibility to ensure protection of this important natural resource is the driving force behind this second edition of *Farmland Preservation*. Since the 2009 release of the first edition of this book, governments and communities have continued to struggle with competing demands for the use of high-quality agricultural land. We now know that climate change will have impacts on agricultural land and productivity around the world, underscoring the urgent need to protect arable land in order to meet the future global demand for food, fibre, and energy. There is also renewed interest in local food, healthy eating, rural resilience, and the contributions of agriculture to local economic and social development.

Individually, several chapters in this book explore issues in specific jurisdictions, such as British Columbia, Quebec, and Australia. These chapters provide in-depth analyses of different approaches at local, provincial (or state), and national levels. Collectively, the chapters present a broad spectrum of thought on issues related to farmland preservation. Presented is a range of ideas and policy options that can assist planners in developing regionally appropriate and effective strategies. Although the book does have a distinctly Canadian focus, it will also be of interest to anyone seeking thoughtful approaches to the preservation of farmland.

Throughout the thirteen chapters in this book, both conceptual contexts of farmland preservation policies and specific examples of jurisdictional approaches to protecting high-quality agricultural land resources are included. The Introduction provides an overview of the different paradigms that shape agricultural land preservation policies, and Chapter 1 provides the historical context of current approaches to these policies. The next four chapters provide insights into different approaches in Quebec, Ontario, and British Columbia.

Evolving issues and creative responses are covered in Chapters 6 through 10. The concept of "Smart Growth" in relation to farmland preservation is discussed within contexts in both Ontario and British Columbia, taking into consideration the complexities of accommodating a growing population on a limited land base. Other issues, such as non-farm severances and farm-level adaptation in the peri-urban space, are also discussed, along with a focused look at a unique response to preserving farmland in Ontario through the use of permanent easements held by the Ontario Farmland Trust.

International perspectives on farmland preservation are provided in the latter chapters and include examples from the United States and Australia. Although the policy and land-use planning frameworks might differ between countries, the issue is universal: how to ensure an adequate and sustainable agricultural land base for future generations in the face of multiple and competing demands. Our hope as editors is that this book will stimulate further discussion, research, and action leading to a renewed commitment to protect farmland for future generations. Although this issue does not always command public attention, it is an important one that awaits strong leadership from planners, farmers, academics, policy makers, and members of society.

INTRODUCTION

Farmland Preservation Perspectives

BRONWYNNE WILTON

Farmland preservation is an issue that societies and governments around the world must face as the global population increases. Current projections are that the global population will reach 9 billion by 2050, and ensuring that an adequate amount of food is produced will continue to challenge the agri-food system. Although feeding the global population requires multiple approaches, such as minimizing food waste, increasing equity in food distribution, and advancing technologies for crop efficiency, a stable and secure base of high-quality agricultural land will be a key component of a resilient and sustainable global food system. In addition to the growing population, the global agricultural system faces uncertainties and challenges because of climate change. Arid regions of the world in particular are facing increased drought conditions, while other regions are experiencing unusual local weather patterns and extreme weather events such as floods and prolonged periods of high temperatures. This global uncertainty places additional pressure on regions such as Canada (in particular Ontario, with its high-quality agricultural lands and favourable climatic conditions) to protect this valuable resource for future generations.

The need to protect farmland from urban development has been recognized in countries around the world (see Bunce 1998; Han and He 1999; Onata et al. 2000; and Yokohari, Brown, and Takeuchi 1994). In fact, over the past three decades, a growing body of literature and policy work related to farmland preservation has been developed in North America in response to increased public concern over the permanent loss of agricultural land (Bunce 1998; Liu and Lynch 2011). Although farmland preservation has been recognized as a priority for jurisdictions around the world, several different protectionist policies have been implemented, depending on national and local involvement in the planning process as well as societal values, ideals, and norms.

This chapter provides an overview of the theories that guide the implementation of preservation policies as well as the varied methods employed by different jurisdictions. A typology of reasons that motivate societies to engage in farmland preservation is presented. A discussion of the unique challenges faced at the urban-rural fringe is also included to provide background on new and innovative approaches being tested in these often highly politicized and controversial regions. Although the focus of this chapter is primarily on North America, and Canada in particular, examples from the literature related to other countries are discussed for the purpose of comparison.

WHAT IS FARMLAND PRESERVATION?

To properly understand farmland preservation, it is important to first understand what exactly is meant by the word *farmland* itself. Agricultural land, or farmland, is any land capable of producing agricultural products, such as grains, livestock feed, tender fruits and vegetables, and pasture for livestock (Ontario, Ministry of Municipal Affairs and Housing 2005). Although definitions of quality farmland vary from one jurisdiction to another, in the Ontario context high-capability agricultural lands are considered to include the following:

- All lands which have a high capability for the production of specialty crops due to special soils or climate.
- All lands where soil classes 1, 2, 3, and 4 predominate as defined in the Canada Land Inventory.
- Additional areas where farms exhibit characteristics of ongoing viable agriculture.
- Additional areas where local market conditions ensure agricultural viability where it might not exist otherwise.
  (Ontario, Ministry of Municipal Affairs and Housing 2005).

The Canada Land Inventory (CLI) classifies land according to its inherent ability to support agriculture. The CLI provides an inventory of all lands in Canada, and there are similar systems in other countries around the world. There are seven classes in the CLI, with Class 1 being the most suitable for agriculture and Class 7 being the least suitable for agriculture because of significant constraints, such as poor soils, unsuitable climate, difficult topography, and excessive stoniness (Ontario, Ministry of Agriculture and Food 1978). Classes 2 and 3 of the CLI have only moderate limitations for common field crops and are often included in the definition of "prime farmland" or "high-quality agricultural lands" (Ontario, Ministry of Agriculture and Food 2001). The CLI is only one example of a land classification system, but it is an important illustration of the use of biophysical information for determining the quality of farmland that requires protection through various farmland preservation policies.

Prior to the 1970s, the predominant public perception was that North America had "a limitless supply of farmland and unbounded technological capabilities" (Bunce 1998, 233). However, throughout the 1960s, evidence of increased soil erosion, degradation, and urban sprawl sparked concern among academics over the loss of productive farmland in North America (Bunce 1998). Throughout the 1970s, numerous studies and policies were implemented to both measure and protect farmland in North America (Ontario Institute of Agrologists 1975; Ontario, Ministry of Agriculture and Food 1977, 1978; United States, Department of Agriculture, and President's Council on Environmental Quality 1981). By the end of the 1970s, many state, provincial, and local jurisdictions had some form of farmland preservation policy in place (Bunce 1998; Furuseth and Pierce 1982).

For example, Rozenbaum and Reganold (1986) discuss the development of ten acts passed in the four-state region of Minnesota, Wisconsin, Illinois, and Iowa. Between 1970 and 1980, this region lost approximately 1 million acres of agricultural land, and policy makers responded with a series of laws to enact different state farmland preservation programs. Similarly, in British Columbia, the rapid conversion of prime farmland was so surprising to both the public and the government that an immediate freeze on development was placed on all agricultural lands in the province. The Provincial Land Commission was developed out of an order-in-council passed by the provincial government in late 1972. Popularly known as the "land freeze," it was passed to prevent the further subdivision and non-farm use of agricultural land that had been occurring at an alarming rate during the 1960s, particularly in the Fraser and Okanagan Valleys (Jackson 1985). This land freeze effectively halted urban development on farmland until long-term statutory provisions could be implemented. In early 1973, a second order-in-council set out the details of the land freeze: "Farmland over two acres in size so classified for taxation purposes, or land designated in Classes 1, 2, 3, or 4 of agricultural capability in [the] Canada Land Inventory," would be set aside in newly established agricultural land reserves (Jackson 1985; also see Chapter 4 of this volume).

As citizens around the globe become increasingly aware of the importance of protecting and managing natural resources in a sustainable manner for a growing population, policy makers are paying greater attention to the preservation of high-quality agricultural land. Historically, land preservation policies developed from rather idealistic or pastoral values, but policy instruments are becoming more restrictive nowadays as the finite nature of this resource becomes more apparent.

COMMON FARMLAND PRESERVATION TOOLS

Farmland preservation policies have taken on different shapes depending, a great deal, on the property rights and land-use laws of the jurisdictions enacting them (Bunce 1998; Feitelson 1999; Rozenbaum and Reganold 1986; Schiffman 1983; Westphal 2001). Westphal (2001) describes property rights as a concept that emanated from the English land tenure system that originated in the Middle Ages. In the United States, for example, property rights give property owners the rights to deed restrictions, liens, mineral resources, easements, use, and development. Although property rights are often thought of as absolute rights, the overall interests of society, especially the health, safety, and welfare of the public, will always overrule individual property rights. This concept is strengthened in Canada, where public policy rather than private property rights guides land use. Table 0.1 provides an overview of common farmland preservation tools with accompanying considerations for implementation.

THE RATIONALE BEHIND FARMLAND PRESERVATION POLICIES

Although farmland preservation is basically about protecting a biophysical resource through various land-use planning policies, the societal rationale behind the policies drives the system. As Feitelson (1999, 432) indicates, "all farmland protection programs are, essentially, a set of conventions and entitlements determining the nature and scope for individual and business choice sets with regard to the use of such land." Feitelson goes on to say that institutional structures of farmland protection policies change over time, "reflecting social norms, power structures, and production and consumption possibilities" (432). Feitelson also lists six rationales for farmland protection policies, including food security, negative externalities of the urban fringe,

Table 0.1. Common farmland preservation tools.

TOOL	JURISDICTION	"STRENGTH" OF TOOL	REASONS FOR IMPLEMENTING	PUBLIC ACCEPTANCE
Land-use planning	Primary method of farmland preservation in Canada	Varies, depending on political support	To encourage the "wise use" of resources for the benefit of society	Widely accepted in Canada; can be viewed as an infringement on property rights
Agricultural land reserves	British Columbia	Very strong but subject to political will	To permanently protect agricultural resources in a specific region (BC)	Strong level of public support
Purchase of development rights (PDR)	Common tool used in the United States	Very strong but dependent on high levels of financial support	To protect specific tracts of land for permanent agricultural use	Strong but applicable only where property right laws allow for the transfer of development rights

Source: Greater Toronto Services Board, *A GTA Countryside Strategy: Draft Strategic Directions*, 2000.

loss of positive externalities, the rural idyll, the agrarian idyll, and a land ethic. Spaling and Wood (1998) describe the rationales that drive farmland protection policies under three main headings: econocentric, biocentric, and theocentric. For his part, Bunce (1998) describes four principal rationales, including resource management (including food security), ecological conservation, local amenity protection, and agrarian ideals. We see these rationales coming together today through the recent review of four growth plans in the Greater Golden Horseshoe area, the area encompassing Toronto and its surrounding metropolitan regions (Ontario, Ministry of Municipal Affairs and Housing 2015).

A FARMLAND PRESERVATION TYPOLOGY

Several different rationales in the literature motivate the development and support of farmland preservation policies. Bunce (1998, 234) explains that the early studies and legislation related to farmland preservation developed out of concern over food security that emerged at the same time as the environmental movement of the 1960s: "Productionist arguments dominated the emerging discourse of the farmland preservation movement. If so much farmland was being converted to non-agricultural uses, it seemed obvious that food production would be threatened." He then asks, "what . . . has sustained farmland preservation as a separate and contentious rural planning issue?" (234). Other motivations for restricting the urbanization of agricultural lands include controlling urban sprawl, preserving countryside amenities, protecting the natural environment, maintaining rural communities and the farming way of life, and, in the case of Quebec, guarding the national identity (Bunce 1998; McCallum 1994). The following summary of the various rationales serves as the foundation for further study and exploration.

Utilitarian Perspective – This rationale is based primarily upon food production and the market value of land. From the utilitarian perspective, buyers and sellers in the market determine the value of land viewed as a commodity, and this has been the dominant perspective of Western society (Spaling and Wood 1998).

Resource Management Rationale – The resource management rationale is also based upon food production; however, it has broader connections to the environmental movement of the 1960s (Bunce 1998). The Malthusian spectre of rapid population growth outstripping food production capacity emerged in the 1960s in North America. From this concern over the misuse of agricultural land, extensive work was done on land classification systems, such as the Canadian Land Inventory. The capabilities and qualities of the land base were emphasized to direct land-use planning in an appropriate manner as much as possible. This rationale formed the early language of the farmland preservation movement—such as "shrinking land base" and "disappearing farmland"—found in so many of the early publications on farmland preservation (Bunce 1998).

Ecological Conservation Perspective – The land capability issues that emerged through the resource management rationale brought the farmland preservation movement under a broader, more ecologically oriented, perspective. The ecological conservation rationale was strongly influenced by Aldo Leopold's *A Sand County Almanac* (1949), in which he wrote about a land ethic, or sense of stewardship toward the land.

This perspective also became more important during the 1960s and 1970s as the food productionist argument weakened and the environmental movement strengthened following the publication of Rachel Carson's *Silent Spring* (1962). Wendell Berry (1977, 1981) also wrote about the need to cherish and protect land for its inherent values as well as human use. This rationale encompasses a broader philosophy of land management based upon the stewardship of natural resources for intrinsic purposes as opposed to strictly anthropocentric needs.

Local Amenity Protection Rationale – The protection of rural amenities and communities has become an increasingly important theme in farmland preservation (Bunce 1998; Moss 2000; Yokohari et al. 1994). This rationale emerged during the 1980s, again as the food productionist argument weakened (Bunce 1998; Feitelson 1999). Amenity protection encompasses rural amenities such as open space, recreation, clean air and water, wildlife habitat, and historical character (Feitelson 1999; Moss 2000; Yokohari et al. 1994). As Feitelson writes, "If appropriately managed, farmlands can provide a range of public goods: ground water recharge, storm water management, water pollution control (including the recycling of wastewater) and enhancement of biodiversity" (Feitelson 1999, 433). In a study of the conversion of agricultural lands in California, Schiffman (1982, 254) describes the costs associated with the loss of agricultural lands, including "the disappearance of aesthetically pleasing countryside and its wildlife habitat; the loss of areas where groundwater can be replenished and air pollution dispersed; the reduced opportunities for suburbanites to know and learn from the followers of a more simple and self-sufficient lifestyle; and the use of increased energy to truck agricultural produce longer distances to consumers."

Agrarian Ideals Perspective – The agrarian idyll is based upon the farmer as the "agent assuring stewardship of the land, on one hand, and as the basis for community life, on the other hand" (Feitelson 1999, 434). With this rationale, the protection of farmland is based upon the cultural ideology of "rural authenticity" or "agricultural idyll" (Bunce 1998; Feitelson 1999).

Collaboration and Innovation at the Urban Fringe Rationale – The agricultural landscape at the urban fringe is generally recognized as an area where conflict over land use intensifies (Schiffman 1982; Spaling and Wood 1998; Sullivan 1994; Westphal 2001; Yokohari et al. 1994). Sullivan (1994, 86) describes the rural-urban fringe as a region that presents a complex challenge to planners and landscape designers, and decisions about the "development of farmland, rangeland, and forests tend to be driven by economic considerations that all too often ignore the non-economic values people have for place." This conflict can lead to highly charged debates over the use of land: "Anywhere an urban-rural interface exists, a battle is going on between forces that wish to develop the land for its real estate (or fair market value) and those who wish to protect and preserve it for its natural resources and agricultural value" (Westphal 2001, 14).

Spaling and Wood (1998, 105) echo this sentiment when they explain that land-use policies in the rural-urban fringe "often result in conflicting opinions among planners, developers, farmers and rural residents" and that the geographic focus of these conflicts is the conversion of agricultural land to non-agricultural uses. The ethical framework presented earlier in this chapter definitely plays a role in the conflict over land use at the rural-urban fringe, where landowners and decision makers have a great deal at stake (e.g., political pressure, substantial financial implications). Personal and societal ethics related to the value of land, both economic (anthropocentric) and non-economic (non-anthropocentric), come into play in the conversion of agricultural land.

Although this chapter has focused on the preservation of agricultural land and the associated rationales that motivate policies of preservation, it has been suggested that societies should emphasize restricting the sizes of cities and pay more attention to the vitality and livability of existing urban centres, which in turn would protect the countryside. In fact, Alterman (1997, 238) suggests that North America should focus on "containing urban growth" rather than on preserving farmland, since doing so would prevent the further wasteful pattern of urban sprawl seen today: "An Urban Containment Movement will focus on improving urban and suburban land utilization through higher densities, infill, use of underground space, and multiple use, and will explain the importance of good land management for future generations."

Ontario responded to this approach by establishing the Greenbelt Plan in 2005 and the Growth Plan in 2006 under the Places to Grow Act (Ontario, Ministry of Municipal Affairs and Housing 2005; Ontario, Ministry of Infrastructure 2006). The protected countryside identified by the Greenbelt Plan consists of an agricultural system and a natural system together with a series of settlement areas. The agricultural system is made up of specialty crop areas, prime agricultural land areas, and rural areas. To complement this protected area within the greenbelt, the Growth Plan directs development to built-up areas and encourages intensification as a means to preserve open spaces and agricultural lands. The plan provides policy direction for the identification of natural systems and prime agricultural areas and the enhancement of conservation of these resources. It is widely acknowledged that, without proper growth management, negative impacts associated with growth such as declining air and water quality and loss of natural resources, including agricultural lands, will continue.

In response to this suggestion, throughout North America there is a growing movement toward "Smart Growth" and "New Urbanism," attempts to plan for more compact and livable urban areas. According to Szold and Carbonell, "To some, smart growth is simply a euphemism for better choices about future development and land use. To others, however, smart growth principles are specifically those that embody viable alternatives to prevailing suburban sprawl" (Szold and Carbonell 2002, 3). The Smart Growth movement is gaining momentum in North America, but it remains to be seen whether it will have a meaningful impact on settlement patterns or simply be empty rhetoric for politicians.

Innovations at the rural-urban fringe are likely to take on multiple forms to represent the various ethical stances and rationales behind the farmland preservation movement. One innovative approach becoming more popular is the concept of a land trust (Schiffman 1983; Watkins, Hilts, and Brockie 2003). Watkins et al. note that "an Ontario farmland trust could perform any number of functions to protect foodland, from education and government lobbying in support of foodland preservation to acquiring farmland and leasing it back to farmers at affordable rates" (Watkins et al. 2003, 10). Because a land trust is a private, locally based, non-profit, tax-exempt corporation, legally empowered to accept and manage land for the purpose of preservation, it completely sidesteps the politics of farmland preservation (Schiffman 1983). Although this might be a positive attribute in the process of protecting land, the separation from a public process might be a detriment to the needs of society, depending again on the values and ethics of the individuals involved in management of the land trust. The rural-urban fringe will likely remain a contentious region since personal agendas will always play roles in the public process of land-use decision making. Lessons from around the world should be employed to guide land-use planning and farmland preservation policies (Alterman 1997). The concept of a farmland trust in Ontario has become a reality and is discussed in further detail in Chapter 11.

CONCLUSION

This chapter has presented an overview of the literature related to farmland preservation policies, primarily those in North America. A brief definition of farmland has been included as well as a summary of the farmland preservation tools available to policy makers. The initial rationale behind farmland preservation policies was the notion of a disappearing resource related to food production and security. However, since food scarcity never occurred, at least in Western markets, concern over the conversion of agricultural land to urban uses became much broader, coinciding with the environmental movement of the 1960s and then encompassing rural amenities on a much broader scale. These latter characteristics are often supported by both anthropocentric and non-anthropocentric value positions and lead to a multiplicity of voices calling for stronger controls over agricultural land, usually for a variety of reasons. Nowhere is this multiplicity more evident than in the fringes around urban centres, where the demands for land to be used for different purposes increase.

REFERENCES

Alterman, R. 1997. "The Challenge of Farmland Preservation: Lessons from a Six-Nation Comparison." *Journal of the American Planning Association* 63 (2): 221–37.

Bunce, M. 1998. "Thirty Years of Farmland Preservation in North America: Discourses and Ideologies of a Movement." *Journal of Rural Studies* 14 (2): 233–47.

Carson, R. 1962. *Silent Spring*. Cambridge, MA: Riverside Press.

Feitelson, E. 1999. "Social Norms, Rationales, and Policies: Reframing Farmland Protection in Israel." *Journal of Rural Studies* 15: 431–46.

Furuseth, O.J., and J.T. Pierce. 1982. "A Comparative Analysis of Farmland Preservation Programmes in North America." *Canadian Geographer* 26 (3): 191–206.

Greater Toronto Services Board. 2000. *Common Farmland Preservation Tools*. Toronto: City of Toronto.

Han, S.S., and C.X. He. 1999. "Diminishing Farmland and Urban Development in China: 1993–1996." *GeoJournal* 49: 257–67.

Jackson, J.N. 1985. "British Columbia and Ontario: Some Comparisons in the Provincial Approach to Safeguarding Agricultural Land." *Ontario Geography* 2: 11–24.

Leopold, A. 1949. *A Sand County Almanac*. Oxford: Oxford University Press.

Liu, X., and L. Lynch. 2011. "Do Agricultural Land Preservation Programs Reduce Farmland Loss? Evidence from a Propensity Score Matching Estimator." *Land Economics* 887 (2): 183–201.

McCallum, C. 1994. "The Effect of Quebec's Agricultural Preservation Law on Agriculture and Rural Land Use Near Sherbrooke." MA thesis, Carleton University.

Moss, M.R. 2000. "Interdisciplinarity, Landscape Ecology, and the 'Transformation of Agricultural Landscapes.'" *Landscape Ecology* 15: 303–11.

Onata, J.J., E. Anderson, B. Peco, and J. Primdahl. 2000. "Agri-Environmental Schemes and the European Agricultural Landscapes: The Role of Indicators as Valuing Tools for Evaluation." *Landscape Ecology* 15: 271–80.

Ontario Institute of Agrologists. 1975. *Foodland: Preservation or Starvation—A Statement on Land Use Policy by the Ontario Institute of Agrologists*. Erin, ON: Herrington Printing and Publishing Company.

Ontario. Ministry of Agriculture and Food. 1977. *Green Paper on Planning for Agriculture: Food Land Guidelines*. Toronto: Queen's Printer.

——. 1978. *Foodland Guidelines: A Policy Statement of the Government of Ontario on Planning for Agriculture*. Toronto: Queen's Printer.

——. 2001. *Guide to Agricultural Land Use*. Publication 824. Toronto: Queen's Printer.

Ontario. Ministry of Infrastructure. 2006. *Growth Plan for the Greater Golden Horseshoe*. Toronto: Queen's Printer.

Ontario. Ministry of Municipal Affairs and Housing. 2005. *Greenbelt Plan*. Toronto: Queen's Printer.

——. 2015. *Planning for Health, Prosperity, and Growth in the Greater Golden Horseshoe: 2015–2041*. Toronto: Queen's Printer.

Rozenbaum, S.J., and J.P. Reganold. 1986. "State Farmland Preservation Programs within the Upper Mississippi River Basin: A Comparison." *Landscape Planning* 12: 315–36.

Schiffman, I. 1983. "Saving California Farmland: The Politics of Preservation." *Landscape Planning* 9: 249–69.

Spaling, H., and J.R. Wood. 1998. "Greed, need or creed? Farmland ethics in the rural-urban fringe." *Land Use Policy* 15 (2): 105–18.

Sullivan, W.C. 1994. "Perceptions of the Rural-Urban Fringe: Citizen Preferences for Natural and Development Settings." *Landscape and Urban Planning* 29: 85–101.

Szold, T.S., and A. Carbonell. 2002. *Smart Growth: Form and Consequences*. Lincoln Institute of Land Policy. Toronto: Webcom.

United States, Department of Agriculture, and President's Council on Environmental Quality. 1981. *National Agricultural Lands Study*. Washington, DC: Government Printing Office.

Watkins, M., S. Hilts, and E. Brockie. 2003. "Protecting Southern Ontario's Farmland: Challenges and Opportunities." Farmland Preservation Research Project Discussion Paper Series. Centre for Land and Water Stewardship, University of Guelph.

Westphal, J.M. 2001. "Managing Agricultural Resources at the Urban-Rural Interface: A Case Study of the Old Mission Peninsula." *Landscape and Urban Planning* 57: 13–24.

Yokohari, M., R.D. Brown, and K. Takeuchi. 1994. "A Framework for the Conservation of Rural Ecological Landscapes in the Urban Fringe Area in Japan." *Landscape and Urban Planning* 29: 103–16.

CHAPTER ONE

Canadian Farmland: A Fluctuating Commodity

MICHAEL TROUGHTON

This chapter presents a summary of the development of agriculture in Canada, which will provide the context within which farmland preservation (more particularly farmland loss) can be placed. The primary emphasis is on the past seventy years. A brief account of the evolution of the farming system will help the reader to better gauge the types and magnitude of change to the farmland base.

Although Canada has been extensively settled for perhaps 12,000 years, agriculture arrived only about 1,500 years ago, after having spread into eastern North America from its Meso-American hearth to what became southern Ontario. Early agriculture was based on maize (corn), but its Aboriginal practitioners played little direct role in defining the agricultural areas and systems that we know today (Wright and Feckau 1987).

AGRICULTURAL SETTLEMENT BETWEEN 1780 AND 1930

Agriculture today is the latest phase of European-derived, farm-based settlement that began in the seventeenth century but was most active throughout the nineteenth century and the first third of the twentieth century. Settlement was based upon various land alienation and survey systems, including the French long lot, seigneurial system, lot and concession townships in the rest of Quebec and Ontario, and township and range sectional system, borrowed from the United States and placed as a checkerboard across the southern prairie provinces.

Settlement progressed from east to west through the various survey systems into lands that seemed to be good for establishing family farms. The survey did not always identify land that could be settled successfully, however, and there were many failures even among the early subsistence farmers. Nevertheless, by the late nineteenth century, in the more favourable lands in eastern Canada, including Prince Edward Island, the Annapolis Valley in Nova Scotia, the upper St. Lawrence Valley and Montreal Plain in Quebec, and much of Ontario south of the Canadian Shield, agrarian systems were in place (Troughton 1982). They were dominated by medium-sized, mixed family farms, which had moved relatively quickly into commercial operations, supplying both domestic and export markets with products such as wheat flour, beef and pork, cheddar cheese, and fruits and vegetables. By the peak year of 1891, there were more than 580,000 farms, containing 19.75 million hectares in eastern Canada, though less than half the land had been improved, and only 38 percent was in crops (see Table 1.1).

In western Canada, despite some early bridgeheads, extensive settlement had to wait until the transfer of Rupert's Land to the new Dominion of Canada was completed in 1870, and it remained slow until completion of the transcontinental railway in the mid-1880s. Between 1891, when western settlement accounted for just 8 percent of Canada's farmland and just 6 percent of farms, and the 1930s, when growth was cut off by the economic depression and severe drought, a huge area was surveyed and settled in farms. Despite fluctuating immigration, and some local failures within the rigid township survey system, 50 million hectares were in farmland by 1936, almost all (98 percent) in the three prairie provinces. By that time, more than 320,000 farms represented 44 percent of the Canadian total, and the extensive western farmland area accounted for over 70 percent of the Canadian total (see Table 1.2). Western expansion had been accompanied by an even more rapid shift than in eastern Canada into a commercially oriented system, based upon the production and export of wheat.

Although there were some attempts to extend farm-based settlement into and north of the Canadian Shield in Ontario and Quebec, and later along the northern prairie margins, the areas fully occupied by 1890 in eastern Canada and by the 1930s in western Canada have represented the primary farmland base. These areas have experienced significant losses and gains of various types.

Farmland Losses and Gains

Canada's farmland area has never been static. Between the late nineteenth century and the Second World War, two highly contrasting types of change took place: huge increases in farmland area in western Canada and significant losses of farmland area in eastern Canada. Much of the land settled in the five eastern provinces was of poor quality and unable to respond to the

Table 1.1. Farm numbers, farmland areas, and cropland areas, eastern Canada, 1891 and 1941.

PROVINCE/ REGION	FARM NUMBERS (000s)			FARMLAND AREAS (000ha)			CROP LAND AREAS (000ha)		
	1891	1941	%	1891	1941	%	1891	1941	%
PEI	15	12	-20	491	566	+15	217	189	-13
NOVA SCOTIA	65	33	-49	2,461	1,545	-37	392	214	-45
NEW BRUNSWICK	41	32	-22	1,809	1,604	-11	412	346	-16
TOTAL	121	77	-36	4,761	3,715	-22	1,021	749	-27

Source: Dominion of Canada (1891, Vol. 2, Table XVI); Dominion Bureau of Statistics (1941, Vol. 8, Part 1, "Agriculture," Tables 28 and 29).

changing nature of farming. These changes included the shift to commercial production based upon inputs such as the first fertilizers and self-propelled machines. The result was widespread farmland abandonment. Between 1891 and 1941, eastern Canada saw a net loss of 170,000 (29 percent) of its farms. During that period, the Maritime provinces experienced a decline of 1 million hectares (22 percent) of farmland (see Table 1.1). Both Quebec and Ontario experienced some growth, mainly associated with northern (clay belt) settlement, which masked losses in southern marginal areas, notably the Canadian Shield. Northern settlement, however, resulted in only minimal amounts of cropland, especially in Ontario.

Most of Canada still had a predominantly rural population well into the twentieth century, with relatively small and compact cities and towns. Thus, with modest exceptions in the vicinities of Montreal and Toronto, farmland loss through conversion to urban uses, the other major type of loss, was not significant before the Second World War.

The 1941 *Canadian Census of Agriculture* reveals a critical watershed in Canadian agriculture. It records the peak number of farms in any one census (732,832)[1] and a farmland total of 70.5 million hectares, of which 22.6 million hectares (32 percent) were in crops (see Table 1.1). Despite larger average farm size in the west (150 hectares) than in the east (50 hectares), the national farming system was remarkably homogeneous in social and economic terms. Total capital values (TCV) and gross farm receipts (GFR) had been affected by the Depression, and had declined by over 25 percent from peaks in the 1920s, but average values of farm TCV, expenses, and GFR per farm and per hectare hardly varied between east and west. Despite some regional specialization, almost all farms were mixed crop and livestock operations, and less than 9 percent of all farm operators were part-time.

CHANGES TO THE CANADIAN AGRICULTURAL AND FOOD SYSTEM, 1940s TO PRESENT

Intense changes to the Canadian agricultural system, to its family farm structure, and to key parts of the farmland base have marked the past seventy years. In assessing these changes, a division can be made between the initial period, from the late 1940s to about 1970, during which the emphasis was on rationalization within an overbuilt farm system, and the period since 1970, which has seen the industrialization of some farms as part of a broad restructuring within an agri-food commodity chain. Accompanying these changes has been intensive urbanization in Canada. The applications and results of these changes have varied widely by region but nationally have included both losses from abandonment and conversion, and significant increases, especially in cropland totals.

It became obvious in the late 1940s that there was a large number of economically non-viable small farms, a situation exacerbated by the increased use of farm machinery and fertilizers. However, the impacts were different in different regions. In the marginal areas (the periphery), which now included northern areas in Quebec and Ontario, the new inputs could not be supported, and many farms fell further behind in terms of low productivity and income generation. This led to a new wave of farm abandonment. In contrast, in areas with better soils and greater capabilities for agriculture, mechanization meant increased land and labour productivity but resulted in active farm amalgamation—fewer but larger farms. The latter rationalization was active in both the lowland areas of central Canada (southern Quebec and Ontario) and across the prairie provinces. However, in the west, physical limitations reducing the yield per hectare resulted in much larger farm units, whereas in central areas farm size increases were more modest since land was used more intensively. The collective impact of mechanization, abandonment, and amalgamation was a huge national reduction in farm numbers of 50 percent between 1941 and 1971. Proportionally, losses were highest in the Maritimes (79 percent) and Quebec (61 percent), but the prairie provinces lost 41 percent from a total that had peaked in 1941 (see Table 1.2). Farm losses in eastern Canada were accompanied by major losses of total farmland (39 percent) and cropland (23 percent). Again the highest rates of loss were in the Maritimes (63 percent), but absolute totals of over 5.5 million hectares of farmland and 1.2 million hectares of cropland were lost in Quebec and Ontario; a significant amount was associated with a precipitous decline in the northern areas of both provinces (Troughton 1983). In contrast, in western Canada, reductions in farm numbers were accompanied by an increase of 6 million hectares (12 percent) in farmland area and intensification that resulted in an increase of 5.6 million hectares (46 percent) in cropland (see Table 1.2).

Before the 1960s, variable farmland capability, though recognized, was not measured on a consistent basis. However, in the early 1960s, the federal government responded to recommendations from the national Resources for Tomorrow Conference (1961) by establishing the Canada Land Inventory (CLI) to assess the productive capabilities of Canada's resource lands (ARDA 1965). The inventory revealed that only 15 percent of Canada had soils with capabilities for agriculture. Another major survey based upon the 1961 census presented indicators of widespread rural social and economic disadvantage, including areas of low farm and farm family income (ARDA 1964). Although the government's agricultural rehabilitation programs were unable to significantly stem losses on the eastern and northern margins, the CLI confirmed the marginal nature of large areas of previously settled and farmed lands that had been or were being abandoned. The CLI also documented the fact that farming occurred on less than 7 percent of national territory and that there was a strong correlation between the much smaller percentage (3.5 percent) of land in the highest agricultural capability classes (1, 2, and 3) and farmed areas that had successfully modernized via mechanization and greater capital intensification. In particular, in central and far western Canada, the CLI indicated the exceptional value to agricultural production of the restricted areas of high-capability land, especially those in Classes 1 and 2, in the Montreal Plain, in Ontario south of the Canadian Shield, and in the Lower Mainland of British Columbia. These minority areas have become the prime focus of efforts at farmland preservation.

Together with agricultural rationalization, based mainly upon mechanization, the post–Second World War decades witnessed a major shift in Canada in relationships between rural and urban populations and land uses. Although the majority of Canada's population was urban by the Second World War, the shift to urban dominance in terms of economic and social development of land was delayed. By the 1950s, however, in the midst of a postwar, urban-industrial, economic boom, with peak rural to urban migration and urban-oriented immigration, the pressures exerted by urban expansion began to affect rural agricultural areas. The expansion took place in an era of urban development based upon the proliferation of low-density residential suburbs and peripheral industrial zones, supported by

18   Farmland Preservation

mass automobile ownership and mobility, which saw cities expand rapidly into the surrounding countryside (Bryant, Russwurm, and McLellan 1982). This exurban expansion resulted in the first major wave of conversion of farmland to non-farm, non-rural uses.

Types of Region and Variable Conditions of Farmland Loss or Gain

In contrast to the relative homogeneity of the farming system in 1940, the postwar rationalization resulted in fragmentation of the system into three zones: the marginal periphery, the agriculturally viable hinterland, and the increasingly active exurban fringe. Each zone exhibited different demographic and land-use circumstances, which resulted in different characteristics of farmland loss or gain (Troughton 1976). This fragmentation of the agricultural ecumene has persisted.

In the periphery, notably the Maritimes, eastern and northern Quebec, and northern and Canadian Shield Ontario, which had already experienced the largest proportional losses, reductions in farm number and farmland and cropland abandonment continued, despite programs aimed at agricultural rehabilitation (the main preoccupation of governments in the 1960s). By far the greatest reductions of farms, farmland, and cropland were in the peripheral regions of eastern Canada, which significantly reduced their proportion of the Canadian agricultural ecumene. In the west, abandonment was generally less marked, though some areas of recent settlement along the northern prairie margins experienced retreats.

In contrast, the hinterland areas of economically viable agriculture, increasingly concentrated nationally in the prairie block, experienced a different type of farmland change. In western Canada, the 1941–71 period of rationalization included significant farmland and cropland expansion (see Table 1.2). As a result, the proportion of farmland in western Canada rose from 72 percent to 82 percent, and cropland from 70 percent to 81 percent, of the national totals, respectively. During this period, too, average farm sizes increased several-fold, reaching 189 and 200 hectares in Saskatchewan and Alberta, respectively.

The changes to farmland in the eastern Canadian hinterland were, on the whole, negative. However, rates of loss in southern Quebec and Ontario were lower than those in the peripheral and fringe zones of the two provinces. Nevertheless, rationalization resulted in moderate losses of both farmland and cropland. Average farm sizes increased but at much lower levels than in the west. However, in the midst of downward pressures on farm incomes, large numbers of farms shifted from full-time to part-time operations. In Ontario, a report on farm incomes, *The Challenge of Abundance* (Special Committee on Farm Income in Ontario 1969), advocated further reductions in farm numbers and anticipated (incorrectly) that part-time farms would be a passing phase. However, the full-time/part-time division has persisted and is characteristic of polarization between types of farm, noted below.

The economic health of farming produced major concerns, which arose because of problems of rehabilitation at the margins, income maintenance versus rising expenses, and fluctuating commodity prices (the so-called cost-price squeeze). In the regions of more viable farming, a major concern developed during the period over farmland within exurban or rural-urban fringe zones. In these expanding areas, losses to conversion had become apparent. Concern over and investigation of farmland losses, and the corollary of farmland preservation, occupied many individuals and agencies, beginning in the late 1950s and reaching an initial peak in the 1970s (Troughton 1995).

Even before the CLI facilitated monitoring of changes by capability class, there were concerns over losses of land around major cities, including disproportionate losses of Classes 1 and 2 land in their environs. As early as the late 1950s, major losses were documented for all the largest Canadian cities, with the rate of loss proportional to size (Crerar 1963). Particular circumstances varied, but the overall effect was a burgeoning demand for land on which to expand the range of urban land uses, each of which could command a higher economic rent in the land market than even the most valuable farmland. The problem was exacerbated by generally weak land-use planning controls, which allowed land speculators to take land out of farm ownership (and most frequently farm use) in advance of development. This *urban shadow* effect (Russwurm 1975) magnified losses, while broader urban influences within the *urban field* and *commuting zones* weakened the distinct, farm-based, rural fabric and increased the potential for farmland conversion (Bryant et al. 1982).

An initial peak of exurban land demand occurred in the decade from 1966 to 1976. It resulted in major losses of high-capability farmland in major urban areas of central Canada, including the metropolitan Montreal area in southwestern Quebec, the Golden Horseshoe area, including the Greater Toronto Area (GTA) and Niagara Peninsula in southern Ontario, and the small but highly productive areas of the Lower Mainland and Fraser Valley, threatened by the expansion of Greater Vancouver. Major threats to nationally significant areas of horticultural production included the Okanagan Valley.

Concerns over conversion losses gained political expression in Quebec, Ontario, and British Columbia (Troughton 1981). Laws were enacted to protect high-capability farmland from alienation and conversion. The first piece of legislation was the BC Land Commission Act (1973), followed in Quebec by the Loi sur la protection du territoire agricole (1978). Although the BC Land Commission had problems maintaining land in farm uses, the Quebec legislation, allied to stricter municipal planning legislation, was relatively successful in protecting prime agricultural land and diverting development to lower-capability zones. In Ontario, legislation was framed but not enacted, and the only safeguard was a weak set of land-use guidelines (Ontario 1978). In each province, the collective case for preservation came up against opposition from those upholding the "rights" of private property owners, the market/speculative value of whose land would be compromised by its preservation as farmland.

AGRICULTURAL RESTRUCTURING AND INDUSTRIALIZATION, 1970s TO PRESENT

The substantial changes during the period of rationalization seemed to suggest that, by the 1970s, agriculture had reached a position of some stability in scale of farm operation, capital intensity and greater specialization. Stability was linked to government measures to rationalize the dairy and poultry sectors, through supply management, and policies of farm income stabilization (Federal Task Force on Agriculture 1969). The late 1970s and 1980s also saw a decline in the exurban land market under a period of less buoyant economic conditions. However, forces already at work resulted in a new round of significant changes, which again affected farms and farmlands. In this case, however, the driving forces were external to the farming sector.

Agribusiness and corporate involvement in the agricultural and food systems in North America date back to the late nineteenth century and early twentieth century, including the consolidation of farm machinery manufacturing and the beginning of processed foods, notably breakfast cereals and canned soups and vegetables. Despite beginnings in major horticultural areas such as California, however, the corporate and farm sectors remained distinct until after the Second World War. Thereafter, in the United States, began a major shift to large-scale, specialized livestock production, supported by surpluses of feed grain and oilseeds (Hart and Mayda 1998).

Table 1.2. Farm numbers, farmland areas, and improved cropland areas in Canada, 1941-71.

PROVINCE/REGION	FARM NUMBERS (000s)			FARMLAND AREAS (000ha)			CROPLAND AREAS (000ha)		
	1941	1971	%	1941	1971	%	1941	1971	%
PEI	12	5	-58	566	313	+45	189	142	-25
NOVA SCOTIA	33	5	-85	1,545	538	-65	214	98	-54
NEW BRUNSWICK	32	6	-81	1,604	542	-66	346	130	-62
MARITIME REGION*	**77**	**16**	**-79**	**3,715**	**1,393**	**-63**	**749**	**370**	**-52**
QUEBEC	155	61	-61	7,310	4,371	-40	2,453	1,755	-29
ONTARIO	178	95	-47	9,060	6,460	-29	3,671	3,179	-13
CENTRAL REGION	**333**	**156**	**-53**	**16,370**	**10,831**	**-34**	**6,124**	**4,934**	**-19**
EASTERN REGION	**410**	**172**	**-58**	**20,085**	**12,250**	**-39**	**6,873**	**5,304**	**-23**
MANITOBA	58	35	-40	6,989	7,693	+10	2,554	3,692	-45
SASKATCHEWAN	139	77	-45	24,266	26,328	+9	7,987	11,064	-39
ALBERTA	100	63	-37	17,514	20,035	+14	4,959	7,727	-56
PRAIRIE REGION	**297**	**175**	**-41**	**48,769**	**54,056**	**+11**	**15,500**	**22,483**	**-45**
BRITISH COLUMBIA	26	18	-31	1,632	2,356	+44	217	442	-104
WESTERN CANADA	**323**	**193**	**-40**	**50,401**	**56,412**	**+12**	**15,717**	**22,925**	**-46**
CANADA	**733**	**365**	**-50**	**70,486**	**68,662**	**-3**	**22,590**	**28,229**	**-25**

Sources: Dominion Bureau of Statistics (1947, Tables 28 and 29); Statistics Canada (1973, Table 30).
* Excludes Newfoundland.

The move was orchestrated and became dominated by large agribusiness companies, which sought to make farm operations part of a vertically and horizontally integrated, consolidated commodity chain. This new industrial structure became well established in the United States by the 1970s and increasingly became part of the Canadian agricultural and food system.

The restructuring of agriculture has taken place increasingly along industrial lines, based upon further economies of scale in each part of the agri-food chain, including the farm production level. In economic terms, there was a shift from attempts to increase production to emphases on lower unit costs and competition to control a relatively inelastic domestic market. Farming continued to face a cost-price squeeze; in order to purchase the inputs necessary to maintain incomes against falling prices, farms became both larger and even more specialized, especially in livestock production.

Given an inelastic market, the result was that production was rapidly concentrated in a relatively small set of large enterprises, the so-called factory farms of huge cattle feedlots and pig and poultry barns. In turn, farm access to the market required links to processing and retail sectors. Production had to increase, via contracts (or even ownership of production facilities), in order to meet the demands and specifications of fast-food and supermarket chains. Farms operating within this structure, therefore, are part of an integrated, or "assembly line," commodity chain (Troughton 2002).

The impacts of restructuring at the farm level have included further reductions in total numbers of farms and polarization between a minority of large-scale, full-time, "industrial" farms and a majority of generally smaller, often part-time farms. Within this structure, the majority of farms wield little economic strength. The impacts of restructuring and industrialization vary across the country by region and zone.

Since 1970, the eastern periphery has seen a reduction in farm numbers and farmlands, but because of the scale of previous losses these latest reductions have been relatively modest (see Table 1.3). Between 1971 and 2001, total farm area in the Atlantic provinces decreased by over 8,000 hectares (47 percent), while farmlands fell by over 320,000 hectares (23 percent). In a reverse trend, however, the cropped area rose by nearly 90,000 hectares (24 percent). Nevertheless, the locally expanded cropland area represents only 1 percent of the national total and only 7 percent of cropland in eastern Canada. No such revival was seen in the other eastern peripheral areas. Agriculture, except in a few pockets, has continued to decline to very low levels in areas of northern Quebec and Ontario. In contrast, some outlying western areas, notably in the Peace River region, have seen land clearance and increased cropping, while farm numbers continue to decline.

Turning to the major hinterland regions, the past forty years have continued regional trends on a smaller scale than in the previous thirty years. In central Canada, it is hard to differentiate between changes in hinterland regions and those in fringe zones. Overall, farm numbers decreased by 32,000 (52 percent) and 43,000 (45 percent) in Quebec and Ontario, respectively. At the same time, farmland areas fell by 24 percent and 20 percent, but cropland areas increased by 7 percent and 14 percent. Cropping intensification was more evident on the larger land base in Ontario, where the cropland area has grown back to a total only marginally below that reached in 1941, albeit on only half of the former number of farms and on 80 percent of the farmland area.

Cropping intensification between 1971 and 2001 was even more pronounced across the west. Farm numbers declined by over 77,000 (40 percent), farmland areas decreased by 2 percent, but cropland areas increased by 28 percent. The past three decades have seen a highly variable farm economic situation, especially affecting western production geared to export markets. Although farmers continue to farm larger areas (with huge increases in the proportion of rented land), they are also diversifying their cash crops (Carlyle 2004) and shifting to intensive livestock production, especially beef cattle and pigs (Ramsey and Everitt 2002). The reduction in farm numbers has been part of a major economic restructuring of rural communities. Grain elevators and delivery points across the Prairies have been reduced by 85 percent and 70 percent, respectively, eliminating several thousand agricultural service communities. These reductions represent a reversal similar in magnitude to the initial rapid growth of the prairie system. They are part of a major reversal in the demographic and community structure of agriculture and a severe threat to the continuation of family farming in the prairie provinces (Troughton 1999).

Finally, turning to the exurban (rural-urban fringe) zone, demand for land for urban and non-farm uses has fluctuated, from relatively low levels during at least two periods of economic recession to very high levels from the late 1990s to the present, especially in the Greater Toronto Area and the Lower Mainland, where development pressures are high. Against these demands has been weak protection by any level of government, especially of farmland per se. Back in the 1970s, concern was expressed in Ontario that, if farmland continued to be lost at the 1966–71 rate, it would be halved by 2000, and the province would lack basic resources for food self-sufficiency (Ontario Institute of Agrologists 1975). Although farmland has continued to be lost, it is being lost at a lower rate, and cropland actually increased by 25 percent between 1971 and 2011. Furthermore, though the per capita availability has fallen, in today's era of increased production per hectare and a global food marketplace, such concerns have become less common. Nevertheless, losses of key blocks of the highest-capability land continue to occur, and a majority of all Class 1 land in Canada lies within 100 kilometres of the edge of Canada's largest and fastest growing metropolitan area.

It might be argued that the demand for land for residential-recreational uses, which finds expression in the proliferation of exurban "hobby farm" properties, can curb total urban conversions. On the one hand, a hobby farm usually means the end of conventional farm production; on the other,

Table 1.3. Farm numbers, farmland areas, and improved cropland areas in Canada, 1971–2011.

PROVINCE/REGION	FARM NUMBERS (000s)			FARMLAND AREAS (000ha)			CROPLAND AREAS (000ha)		
	1971	2011	%	1971	2011	%	1971	2011	%
NEWFOUNDLAND AND LABRADOR	1	0.5	-50	25	31	+24	4	8	+100
PEI	5	1.5	-60	313	241	-23	142	166	+17
NOVA SCOTIA	5	4	-20	538	412	-23	98	114	+16
NEW BRUNSWICK	6	3	-50	542	380	-30	130	142	+9
MARITIME REGION	**17**	**9**	**-47**	**1,418**	**1,064**	**-25**	**374**	**430**	**+18**
QUEBEC	61	29	-52	4,371	3,341	-24	1,755	1,875	+7
ONTARIO	95	52	-45	6,460	5,127	-20	3,179	3,614	+14
CENTRAL REGION	**156**	**81**	**-48**	**10,831**	**8,468**	**-22**	**4,934**	**5,489**	**+11**
EASTERN REGION	**173**	**90**	**-48**	**12,249**	**9,532**	**-22**	**5,308**	**5,919**	**+12**
MANITOBA	35	16	-54	7,693	7,294	-5	3,692	4,349	+18
SASKATCHEWAN	77	37	-52	26,328	24,941	-5	11,064	14,729	+33
ALBERTA	63	43	-32	20,035	20,437	+2	7,727	9,754	+26
PRAIRIE REGION	**175**	**96**	**-45**	**54,056**	**52,672**	**-2**	**22,483**	**28,832**	**+28**
BRITISH COLUMBIA	18	20	+11	2,356	2,611	+11	442	600	+36
WESTERN CANADA	**193**	**116**	**-40**	**56,412**	**55,283**	**-2**	**22,925**	**29,432**	**+28**
CANADA	**366**	**206**	**-44**	**68,661**	**64,815**	**-5**	**28,233**	**35,351**	**+25**

Sources: Statistics Canada (1973, Table 30; 2011)

it can be considered part of a city's open space amenity, prompting broader concerns over its actual and potential loss. Recognition of farmland as a key part of the rural landscape is prompting some re-evaluation of traditional approaches to urban growth, and some individual landowners and environmental organizations are setting land aside, including some farmland. However, in the areas of greatest threat to prime farmland, the pressures of development make land preservation very difficult.

CONCLUSION

In the broader context of the national agricultural system, the idea of farmland preservation presents a paradox. At some times, and in particular places, farmland loss has been a critical concern, with losses assessed as significant in a national context, especially reductions of small areas of high-capability (Class 1) farmland. Yet, since 1941, though farm numbers nationally have declined by 71 percent, the farmland base has experienced a net loss of only 5.6 million hectares (or 8 percent), and the cropland area has expanded by 12.8 million hectares (or 56 percent). Despite losses of some of the highest-capability land, overall increases in yield have more than compensated in terms of production.

It is arguable, however, that the really important changes have been in farm numbers, not only the overall reduction of 71 percent since 1941 but also the shift to a small minority of farms, which in 2001 produced 83 percent of GFR and controlled 59 percent of farm capital, versus the large majority (over 70 percent), which accounted for only 17 percent of GFR while owning 41 percent of TCV. The small minority of "industrial" farms is using less than half the farmland and cropland areas to produce the overwhelming amount of output, including most livestock. In contrast, many of the low-output majority group, which can be considered as having more of a stewardship role, are struggling to supplement low GFRs and retain acceptable family farm incomes from off-farm employment. Evidence suggests that only a minority of such farm families will continue to farm under present conditions. Thus, it is arguable that farmland preservation, in real terms, might be a function of whether a broader basis of economically viable family farming can be recreated.

NOTES

1   The total of all farms in the census records for Canada is achieved by summing up the maximum number for each province, regardless of the date. This method yields a "maximum" of 907,721, but the maximum in a given census is 732,832 in 1941.

REFERENCES

ARDA (Agricultural and Rural Development Act). 1964. *Economic and Social Disadvantage in Canada: Some Graphic Indicators of Location and Degree* (map folio). Ottawa: Department of Forestry.

———. 1965. *The Canada Land Inventory: Objectives, Scope, and Organization*. Ottawa: Department of Forestry.

Bryant, C.R., L.H. Russwurm, and A.G. McLellan. 1982. *The City's Countryside*. Harlow, UK: Longman.

Carlyle, W.J. 2004. "The Rise of Speciality Crops in Saskatchewan, 1981–2001." *Canadian Geographer* 48 (2): 137–51.

Crerar, A.D. 1963. "The Loss of Farmland in the Growth of Metropolitan Regions of Canada." In *Resources for Tomorrow: Supplementary Volume*, 181–96. Ottawa: Queen's Printer.

Dominion Bureau of Statistics. 1947. *Eighth Census of Canada 1941*, vol. 8, part 1, "Agriculture," Tables 28 and 29. Ottawa: King's Printer.

Dominion of Canada. 1892. *Census of Canada 1891*, vol. 2, Table XVI.

Federal Task Force on Agriculture. 1969. *Canadian Agriculture in the Seventies*. Ottawa: Queen's Printer.

Hart, J.F., and C. Mayda. 1998. "The Industrialization of Livestock Production in the United States." *Southeastern Geographer* 33 (1): 58–78.

Ontario Institute of Agrologists. 1975. *Foodland: Preservation or Starvation*. Toronto: Ontario Institute of Agrologists.

Ontario. Ministry of Agriculture and Food. 1978. *Food Land Guidelines: A Policy Statement of the Government of Ontario on Planning for Agriculture*. Toronto: Government of Ontario.

Ramsey, D., and J.C. Everitt. 2002. "Post-Crow Farming in Manitoba." In *Writing Off the Rural West*, edited by R. Epp and D. Whitson, 3–20. Edmonton: University of Alberta Press and Parkland Institute.

Russwurm, L.H. 1975. "Urban Fringe and Urban Shadow." In *Urban Problems*, edited by R. Bryfogle and R.R. Krueger, 148–64. Toronto: Holt, Rinehart, and Winston.

Special Committee on Farm Income in Ontario. 1969. *The Challenge of Abundance*. Toronto: Special Committee on Farm Income in Ontario.

Statistics Canada. 1973. *Census of Canada 1971*. Vol. 1, part 1, "Agriculture," Table 30. Ottawa: Information Canada.

———. 2011. *Census of Agriculture: Snapshot of Canadian Agriculture*. http://www.statcan.gc.ca/pub/95-640-x/2011001/p1/p1-00-eng.htm.

Troughton, M.J. 1976. "A Model for Rural Settlement in Canada." *International Geography* 6: 339–45.

———. 1981. "The Policy and Legislative Response to Loss of Agricultural Land in Canada." *Ontario Geography* 18: 79–109.

———. 1982. *Canadian Agriculture*. Budapest: Akadeniai Kiado.

———. 1983. "The Failure of Agricultural Settlement in Northern Ontario." *Nordia* 17 (1): 141–51.

———. 1995. "Rural Canada and Canadian Rural Geography: An Appraisal." *Canadian Geographer* 39 (4): 290–305.

———. 1999. "Failure, Abandonment, Obsolescence, and Loss of Rural Heritage Elements vs. the Needs of Sustainable Rural Systems." In *Reshaping Rural Ecologies, Economics, and Communities*, edited by J. Pierce, S. Prager, and R. Smith, 201–09. Burnaby, BC: Simon Fraser University.

———. 2002. "Enterprises and Commodity Chains." In *The Sustainability of Rural Systems: Geographical Interpretations*, edited by I. Bowler, C. Bryant, and C. Cocklin, 123–45. Dordrecht: Kluwer Academic Publishers.

Wright, J.V., and R. Feckau. 1987. "Iroquoian Agricultural Settlement." In *Historical Atlas of Canada, Volume 1: From the Beginning to 1800*, edited by R.C. Harris and G. Matthews, plate 12. Toronto: University of Toronto Press.

CHAPTER TWO

Agricultural Land Protection in Quebec

CHRISTOPHER BRYANT, CLAUDE MAROIS, DENIS GRANJON, AND GHALIA CHAHINE

We start this chapter with a simple and probably self-evident proposition for people directly involved in managing land resources: any human activity that takes place in an area or territory is "socially" constructed. A territory is constructed through the actions of different actors as they pursue their own interests, which can include some collective interests. Collective actors include different levels of government, farm organizations, and individual actors such as farmers, who might espouse certain collective values (Giddens 1984; Gumuchian et al. 2003; Marsden et al. 1993).

Once this is recognized, farmland protection and the development of viable agricultural activities can also be seen as being socially constructed. This focus is the result of government intervention in the discharge of duties regarding land-use planning and the actions of the farmers who manage the land, as well as many other actors, including consumer groups who have an interest in farmland as a resource, along with its products and services (Bryant 2012a; Furuseth and Lapping 2001).

AN ANALYTIC FRAMEWORK

At the Université de Montréal since the mid-1990s, a major research theme has been the dynamic of rural localities: how rural areas evolve and the roles of local actors in their construction (e.g., Bryant 1995a, 1995b; Bryant, Desroches, and Juneau 1998; Frej et al. 2003). A conceptual model has been developed based upon seven components. Briefly, to understand the development of a particular territory, it is important first to identify the actors (1) and their interests, values, and objectives (2); then actors pursue their interests and objectives by taking action (or by not taking action) (3) and frequently by mobilizing the resources in the networks in which they operate (4). These networks reflect formal organizational structures, such as the hierarchical structures of government, farmer associations, and syndicates, as well as informal organizational structures, such as friends, family members, colleagues, and political networks (5a, 5b). The cumulative effect of the actions of all the actors in an area or territory gives rise to a territorial identity or profile, characterized by a specific set of orientations (e.g., intensive agriculture, a strong component of farmland conservation, a strong presence of community-supported agriculture, or rural tourism) (6a). Some orientations might remain latent (e.g., a movement to develop a viable farm community or agriculture oriented to direct marketing) (6b). All of these interactions and actions take place in different contexts and at different scales (7). At the provincial level, this includes the legislative context, the economic context in which agriculture finds itself, the cultural context, and the acceptability of public intervention to protect farmland.

In this chapter, we deal principally with three components:

1. the context, particularly in relation to land-use planning;
2. the key actors (in relation to farmland protection in Quebec); and
3. the actions in the construction of localities in the peri-urban zone, illustrated with a number of examples later in the chapter.

At the provincial level, politics, the economic profile of agriculture, and legislation all influence farmland protection. Regarding the political dimension, farmland protection in Quebec received a tremendous boost in the mid-1970s as it became part of the platform of the political program of the Parti Québécois. Protecting farmland was not just an issue of protecting agricultural production capacity from encroachment by urban development but also an important ingredient in feeding a new country should Quebec become independent. Agriculture in Quebec came out of the 1960s behind the main agricultural regions in Canada in terms of technology and modernization (Bryant 2011, 2012b), and significant investment was made to modernize the sector with considerable public sector help, ultimately giving rise to a relatively indebted agricultural structure. This economic situation in many ways has plagued Quebec's agriculture ever since and exacerbates fragile economic structures in several regions. All of this helps to account for the development of the second real farmland protection legislative framework and program in Canada in 1978, the Loi sur la protection du territoire agricole du Québec (LPTAQ).

The principal players in our discussion are

1. the farmers themselves, some of whom are ambiguous about farmland protection since it restricts their ability to capitalize on the non-farm values of their properties, a phenomenon noted in many other jurisdictions (Bryant and Russwurm 1979);

2. the Union des Producteurs Agricoles du Québec (UPA, the Quebec Farmers' Union), which early on defended farmers' right to farm and, through local branches, is often (and increasingly) involved in discussions about farmland protection and planning (UPA 2003);

3. the Commission de la Protection du Territoire Agricole du Québéc (CPTAQ) (the Farmland Protection Commission for Quebec), which oversees implementation of the farmland protection legislation (CPTAQ 2007);

4. the Municipalités Régionales de Comté (MRCs) (municipal regional counties) and the Agglomeration of Montreal (also labelled an MRC), which develop regional land-use plans, with which their constituent local municipalities must conform, as well as municipalities after 1997, when the 1978 law was modified to allow increased involvement of municipalities and a greater degree of integration between the protection of agricultural activities and land-use planning; and

5. the Communauté Métropolitaine de Montréal (CMM) (Montreal Metropolitan Community), whose mission includes providing a regional-level plan (CMM 2012; RMM 2001) that integrates feedback from the provincial government (Québec 1994, 1997), reflections by the CMM itself, and public consultations organized by the CMM (CMM 2012).

THE LEGISLATIVE FRAMEWORK

In 1978, Quebec adopted legislation to protect agricultural land from urban development with the LPTAQ, following in the footsteps of legislation in British Columbia five years earlier. The farmland protection legislation, overseen by the CPTAQ, covers all of Quebec south of fifty degrees latitude. While respecting acquired rights, the commission regulates land for uses other than agriculture, property subdivision, the cutting down of maples in maple stands, the removal of topsoil, and the acquisition of property by non-residents. Some changes were made to the law in 1997 to allow for better integration with land-use planning; important, given that the LPTAQ is in charge of evaluating all exclusion requests made by MRCs.

The initial legislative framework was put in place more or less at the same time as legislation on land-use and urban planning (Loi sur l'aménagement et l'urbanisme [LAU]), an important component of which was the creation of MRCs. The latter were actually instituted over the period 1980-82.

Early assessments of the farmland protection program were positive (see Thibodeau 1984; Thibodeau, Gaudreau, and Bergeron 1986), though in the period covered by these assessments pressures for urban development were not particularly strong. In any case, it is difficult to assess farmland protection programs on the basis of inclusions in and exclusions from the agricultural zones over a particular period of time, since one would need to know exactly what the criteria were in initially delimiting the zones and how realistic the plans were for anticipating urban growth. For instance, the fact that within the boundaries of the CMM 27,654 hectares were removed from the agricultural zone for urban development projects between 1981 and 1996 (Tanguay and Arpin 2001) is difficult to interpret as either positive or negative—it all depends on how adequate the initial areas identified for urban development zones were and how reasonable the delimitation of agricultural zones was. Similar questions can be asked of more recent research on exclusions (Marois 2010; Montminy 2010). The challenges of governance of agricultural territories in peri-urban areas around Montreal remain significant (Marois 2008) and extend beyond exclusions from agricultural zones to include other land-use planning indicators, such as permitted density of development and the fact that there were almost 19,000 hectares available in zones identified for construction.

The adequacy and consistency of decisions of the CPTAQ have also been questioned from time to time (e.g., the cases studied by Deslauriers 1995). More importantly, the Quebec initiative initially suffered from a lack of attention to farming as a socio-economic activity from which farmers and their families had to earn a reasonable living. Some of these criticisms were partially remedied in modifications to the framework in the late

1990s, referred to earlier. In particular, under the Loi modifiant la Loi sur la protection du territoire agricole et d'autres dispositions législatives afin de favoriser la protection des activités agricoles, the commission was

1. charged with taking the activity of farming more into account; the intent was that farmland protection should be accompanied by other policies and measures to support farming in peri-urban areas; and

2. required to take into account the choices made by the MRCs in the revision of their own regional land-use plans, implying better planning for both urban and agricultural zones.

Furthermore, this modification to the original legislation provided for the creation of Local Agricultural Consultative Committees (Comité Consultatif Agricole [CCA]) in the MRCs and other municipal and supramunicipal structures. This has given the agricultural zone and farming much higher profiles in MRC regional land-use planning and provided a vehicle for the farming community to become more involved (CPTAQ 2007).

A major change occurred in 2008, when the province launched a project to support the construction of Development Plans for Agriculture (PDZA) for the agricultural lands in agricultural reserves. This project started with eight MRCs and was subsequently extended to other MRCs with agricultural reserves. In the development of these plans, there was a strong suggestion by the province (Ministry of Agriculture) to take account of the multi-functionality of farmland and implicate various non-farm actors (MAPAQ 2012a, MAPAQ 2012b). This was all related to the need to reinforce farmland preservation by trying to ensure that farms in the agricultural reserves were financially sustainable. Since 2008, what has been happening in Quebec is extremely innovative as the province recognizes that farmland conservation cannot simply be guaranteed through land-use planning and farmland preservation legislation. All factors that affect agriculture must be taken into account in a development plan (akin to a strategic development plan for agriculture). For example, in September 2016, a pilot project was put in place for a particular MRC to see how agricultural adaptation to climate change could be integrated into the PDZA, with the intention of ultimately encouraging this in other MRCs.

Early in the twenty-first century, another criticism was levelled at the CPTAQ (Dumoulin and Marois 2003), specifically in terms of the commission's apparent lack of flexibility in its definitions of "agriculture" and "farming activity." One interpretation is that the commission had not yet fully understood that the multifunctionality of agriculture in urban or peri-urban areas can help to protect farmland and contribute to the development of agricultural activity in those areas (e.g., agri-tourism, direct farm sales, and promotion of farming in maintaining attractive natural and cultural landscapes) (Bryant, Russwurm, and McLellan 1982). This indecision remains today and is especially visible in agricultural peri-urban areas where there has been no provision for nuances, such as smaller lot sizes and innovative techniques on lower-grade lands.

The importance of these other functions for protecting farmland is twofold:

(1) some of these other functions can bring in supplemental farm income, thus strengthening agricultural activity; and

(2) they all involve recognition of urban interest in farmland and farming, thus contributing to political support for the protection of farmland.

As long as the importance of these other functions for protecting farmland is not fully appreciated or understood, there will always be a significant challenge to any provincial-level program both for protecting farmland and for ensuring the maintenance and development of agricultural activity in reserved agricultural zones.

FARMING ZONES AND DESIRED COMBINATIONS OF AGRICULTURAL AND OTHER FUNCTIONS OF FARMLAND

Land-use planning, including farmland protection programs, is not in itself sufficient to protect farmland. As the Ontario Greenbelt Task Force emphasized, "land-use planning alone is insufficient to ensure that agricultural lands within the greenbelt will be farmed" (Ontario 2004, 15).

Land-use planning needs to be complemented by other policies and measures (Bunce 2008). This is easier said than done because peri-urban areas are heterogeneous: what might be appropriate for one area might not necessarily be appropriate for another. This strongly suggests that maintenance and development of farming activity as parts of a policy of protecting farmland require different types of intervention at different levels—a truly management approach to protecting farmland (Ontario 2004).

With this in mind, we propose Figure 2.1. Peri-urban agricultural space is heterogeneous in terms of its structure and dynamic. One simple framework proposed by Bryant (1984) involved identifying three types of agricultural landscape in terms of their dynamics (of course, if we introduce different types of farming [structures and systems], then we can multiply the number of agricultural landscape dynamics substantially). The three broad types suggested were as follows:

1. **Degenerating agricultural landscapes.** They are characterized by all the indicators of agriculture in decline in peri-urban areas, such as idle farmland, speculative land prices, fragmented farmlands and farms, and disinvestment in farms; these difficult circumstances can be exacerbated, of course, by other conditions unfavourable for agricultural production, such as poor market conditions for produce.

2. **Adapting agricultural landscapes.** In these types of landscapes, a significant number of farmers have coped with the stresses in their decision-making environment (e.g., urban development pressures) and adapted, for instance by modifying their farming systems to take advantage of opportunities offered by the nearby urban market. Adaptation involves not only "coping" behaviour but also proactive behaviour in identifying and even creating new opportunities and ways of doing business.

3. **"Normally" developing agricultural landscapes.** In these types of zones, no significant urban development pressures exist, and farming appears to follow its "normal" trajectory for the types of agricultural system in the area.

The importance of identifying these different situations is that, depending on the future collectively desired (by communities, cities, the province), different management strategies and packages are necessary. In Figure 2.1, three combinations of function are used:

1. **Agricultural production.** This is the primary function of farmland, to the exclusion of other functions, except perhaps for a secondary landscape support function.

2. **Landscape protection and farming activity.** This function is paramount, but farming activity is required to maintain the landscape.
3. **Recreation/leisure and farming activity.** Agricultural production is combined with support for recreation (e.g., trails) and leisure (e.g., pick-your-own farm sales, farm visits, and other agri-tourism activities).

Other combinations can be added, but these serve the purposes of this chapter. None of these combinations of function involves farming simply as a means of landscape maintenance. Farming remains at the heart of all these desired agricultural dynamics.

Implicit in the development of management strategies to achieve a particular mix of functions is the desire, in many cases, to move the farming landscape into a new category. Generally, for instance, if one starts with a degenerating agricultural landscape, achieving a particular mix of functions implies reversing the degeneration in order to support an adapting or "normally" evolving farm structure and landscape.

Figure 2.1. Analytic grid linking types of agricultural dynamic, desired mix of functions, and desired types of agricultural dynamic.

TYPOLOGY OF INITIAL AGRICULTURAL DYNAMIC	DESIRED MIX OF FUNCTIONS (EXAMPLES)			DESIRED AGRICULTURAL DYNAMIC FOR THE EXAMPLES
	AGRICULTURAL PRODUCTION	LANDSCAPE PROTECTION AND FARMING ACTIVITY	RECREATION/ LEISURE AND FARMING ACTIVITY	
DEGENERATING		LONGUEUIL ⟶ ST. HILAIRE ⟶		ADAPTING
ADAPTING	LAVAL ⟶	ST. HILAIRE ⟶	SENNEVILLE ⟶	ADAPTING AND DEVELOPING NORMALLY
NORMAL	ST. JOSEPH DU LAC ⟶			ADAPTING AND DEVELOPING NORMALLY

Sources: Bryant (1984, 2011); Bryant and Johnston (1992); Bryant et al. (1998); Granjon (2004).
Note: The path identified for Senneville has been altered to emphasize that the "project" initially began in order to guarantee a long-term future for agricultural activity and quickly became a cornerstone for supporting the conservation of landscapes (and their underlying natural environments) and the recreational and leisure functions of farmland and surrounding "green" land.

How does one design and implement appropriate management strategies and packages for a given peri-urban area? In addition to the involvement of non-local actors (e.g., the province, CPTAQ, UPA, CMM), it is essential to involve local actors, including farmers and representatives of the MRCs or other municipal structures, citizens' groups, NGOs, and local development agencies. These local actors are directly involved in the farmland resource in one way or another; they can thus be seen as "constructing their space"—a protected farmland with a maintained and developing agricultural structure—"in context." Important aspects of context were introduced earlier; one of the newest aspects for local actors in the Montreal region is the CMM, part of whose mandate is to work with local and regional actors to maintain and develop agriculture in peri-urban areas (CMM 2012a).

Not all local initiatives to protect farmland and maintain and develop agricultural activity have met with success. The relatively inflexible interpretations of agricultural activities and spaces used by the commission have been suggested as one reason for some of these failures (Dumoulin and Marois 2003). All of the relatively successful examples presented below are in Montreal's peri-urban zone.

LOCAL INITIATIVES TO PROTECT FARMLAND AND MAINTAIN AND DEVELOP AGRICULTURAL ACTIVITY

In a number of instances, local initiatives have developed to accompany and build a viable agricultural activity in a peri-urban area. These initiatives, generally considered to be successful (or becoming successful), involve municipalities and their planning departments as well as other organizations with more interest in developing agricultural activity, such as local representatives of the UPA. Of course, the particular mix of actors involved can vary given the particular dynamic in each territory. In the areas presented below, local planning departments and other development organizations are attempting to develop or rebuild agricultural activities in agricultural zones by working with representatives of the farming profession and integrating such zones into the urban or metropolitan region framework by connecting initiatives to other functions of agricultural space, such as leisure-oriented activities linked to farming and the positive aspects of farm landscapes in a peri-urban environment. These examples were initiated long before the Plan Métropolitain d'Aménagement et de Développement was adopted in 2011 (CMM 2012b), emphasizing the fact that development itself does not need to wait until formal land-use planning frameworks are in place; frequently, in fact, it cannot wait for this to happen. Most initiatives can take many years to complete (from five to ten years and more in the examples below), and this is part of the difficulty of finding local champions to encourage and accompany such projects, particularly considering the potential municipal turnover of councillors every four years.

Longueuil

The City of Longueuil is an excellent representative case (Planchenault 2001, 2008). Located on the south shore of the St. Lawrence River, this urban municipality still possesses some agricultural land that has been the object of a significant planning and development effort since the late 1990s. The zone was in an extremely vulnerable state at that time despite being protected under the LPTAQ, including large areas of idle and underused farmland. The threat of residential expansion had been present for a long time. However, though the area concerned was part of MRC Champlain (before the municipal mergers), a number of players came together to try to bring the area back into productive farming as an integral part of the MRC. Subsequently, the "new" City of Longueuil carried on with the same process. The vision for the agricultural zone was based upon the notion of constructing a city-country continuum with an intensively managed set of land-use planning and other development tools. This would include providing easy access to the zone for urban citizens wishing to make direct purchases of farm produce in it and providing proper signs and spaces for parking. It was thus firmly based upon the idea that the agricultural zone would be a valued component of the city because of the multiple functions that it could support: farm production, leisure activities (e.g., visits to the zone to purchase farm produce), and a landscape component. To achieve this, a continuously adapting agricultural structure is necessary. Finally, farmers were involved in this project from the beginning together with a representative from the city, and it was subsequently envisaged that farmers would continue to be involved in management of the zone, building upon the involvement of farm representatives in the Local Agricultural Consultative Committee. The project encountered some difficulties related to environmental rules that made it difficult to bring back into cultivation farmland that had been abandoned despite having support from the city, the farmers, the Ministry of Municipal Affairs, and the Ministry of Agriculture. However, after almost five years of negotiation, the project is back on track with the development of an experimental Agenda 21 project as the basis (Charbonneau 2010).

St. Hilaire

The Town of St. Hilaire, with a major focus on Mont St. Hilaire, one of the small, "mountain-like" features on the south bank of the St. Lawrence River, and located some thirty-five kilometres southeast of Montreal, has a relatively small area of apple orchards. With a population of 18,700 in 2011, St. Hilaire is a small but vibrant community; it has a strong service centre function. It has experienced considerable population growth because of its proximity to Montreal, and this connection has been encouraged by the recent opening of a train station in the town as part of the suburban network. Mont St. Hilaire is owned by McGill University (the Gault natural reserve). There is a research facility on part of the mountain and a nature centre to facilitate interpretation for visitors, who also appreciate the trails on the mountain. It is sufficiently special that it was designated as a biosphere reserve by UNESCO in 1978. The orchard area on the lower southern slope of the mountain is relatively small, and only from twelve to fifteen farmers manage it (Barreteau 1997; Granjon 2004). The farms are small, and most of the farmers are part time and have other sources of income. Apple production generally is precarious economically, and this area is no exception. However, many of the orchard farmers in this zone have added some form of agri-tourism, such as pick-your-own sales and small eating areas and/or stores where customers can purchase products derived from the apples—pies, butter, and so on. In addition to what the farmers have created in terms of economic multifunctionality, the orchards are important to several other local actors: the non-farm residents who find that the visual qualities of the orchards add to the local environment (e.g., the apple blossoms in the spring), the nature centre on the mountain, which benefits from visitors to the orchards, the town itself, and MRC Vallée du Richelieu.

For a long time, the orchard area was more or less left to itself, and the farmers themselves remained unorganized. Early on, the town did place signs with directions to the main orchards, but part of the rationale for doing so was to try to manage the heavy traffic at certain times of the year, such as the apple blossom period, to reduce incompatibilities with neighbouring non-farm residents. Subsequently, the farmers organized themselves into an association, and the town and MRC became more involved, using a Local Agricultural Consultative Committee to bring the players together; the Centre Local de Développement (CLD, Local Development Centre) and local representative of the UPA have also become involved. The CLD and MRC see the orchards as an important component of the region's potential for tourism. Finally, the farmers' association has embarked since

2002 on the development of a strategic planning process for the orchard area, including marketing and market development.

The orchard area in St. Hilaire overlaps the first two categories of agricultural dynamics presented in Figure 2.1: that is, degenerating and adapting agricultural landscapes. The actions are aimed at ensuring that adaptation continues and that eventually a combination of adaptation and "normal" agricultural development can be pursued to guarantee survival of the farmland.

St. Joseph du Lac

This village had a population of 6,915 in 2011 and is part of MRC de Deux-Montagnes. Located northwest of the Montreal agglomeration, it is only about thirty minutes from Montreal but has not yet seen the same degree of residential development as St. Hilaire (Granjon 2004). The orchard area is more clearly demarcated from the residential and commercial areas of the municipality, and many of the approximately seventy orchard farmers had already recognized the advantages of their proximity to the urban market and adapted to it. Various types of agri-tourism have been developed on several of the farms, including kiosks and small stores selling apples, products derived from apples, and pick-your-own activities. Local actors other than the farmers have seen the potential of this resource in terms of tourism, including the Office du Tourisme Basses-Laurentides and Corporation Pommes en Fête, established in 1991. The municipality itself became involved more recently but not in a particularly proactive manner. Nonetheless, the agricultural zone is protected under the LPTAQ, and its future has been taken in hand by a variety of local and regional actors, including, of course, the farmers themselves.

In the case of St. Joseph du Lac, the orchard area for a long time could be classed as an agricultural area undergoing "normal" development; subsequently, the farmers became involved in adapting to the nearby urban market. Given current developments, the orchard area is set on a path of continuous adaptation, tending toward "normal" patterns of development for the type of farm structure present.

Laval

The City of Laval (coincident with MRC Laval), immediately north of the Island of Montreal, was the second-largest city in Quebec until the series of municipal mergers in 2002; it was created in 1965 from the merger of fourteen municipalities and had a population at the time of 170,000. It was anticipated at that time that the population of the new city would rise by 2,000 to 650,000.

Its population in 2001 was only 343,005, however, and 401,533 in 2011. This shortfall provides the backdrop to why, after the LPTAQ came into force, the city—working with other actors—decided to reserve a significant part of its area for agriculture and embark on an effort to develop agriculture and make it a cornerstone of Laval's image (Darly 2001). Initially in 1979, the CPTAQ had identified 55 percent of the area of the island on which the City of Laval is situated as part of the agricultural zone, but with negotiations among the CPTAQ, the local branch of the UPA, and the city, the amount was ultimately set at 33 percent of the island's area (1989). In 2001, the City of Laval still included about 200 farms.

Since the agricultural zone is still seen as an economic space, its horticultural activities (flowers, nurseries, market gardens) have also been placed at the base of an institutional structure involving research and training facilities. Furthermore, agriculture has a relatively high profile in the city through AGRIL (Agriculture Laval) and Laval Technopôle, defined by five axes, one of which is the Agropôle. The latter has been involved in a number of projects, including the development of various institutes for training and research in horticulture, and it acts as a vehicle through which dialogue is encouraged with the farming community. One project that relates directly to the concept of multifunctionality is the Route des Fleurs (the Flower Route), in which the Table de Concertation Agroalimentaire de Laval (the Laval Agri-Food Working Group) and, naturally, representatives of the local farming community have been involved. This project, like others with similar orientations, aims to bring the farming community into the city as fully fledged partners in the development of the city, and it recognizes the different contributions of this intensive farming area. All of these activities have turned the city into "the horticultural capital of Quebec" (City of Laval 2012). Although most of the farming community and several local actors directly involved in farming see Laval's agriculture as primarily an economic activity that has followed a "normal" path of development for intensive agriculture, the efforts to incorporate more agri-tourism activities into some farms are taking Laval's farm area more toward a multifunctional agricultural area. The emphasis is on following "normal" patterns of change (investment, structure) for the specialized farms there and continuously adapting to changing circumstances and proximity of the urban population.

Finally, the agricultural zone has been shrinking, and much of the western part of the City of Laval has experienced strong speculation. Although the initiatives to support agriculture noted above have not been as vigorously supported in the past ten to fifteen years, it is once again becoming rejuvenated through the process of constructing a Plan de Développement de la Zone Agricole (PDZA), an initiative of the Ministry of Agriculture in MRCs, together with the active engagement of local actors, including municipalities.

Senneville

Senneville is the western end of the Island of Montreal. The Senneville project started in 2009 to use appropriation of the non-agricultural functions of farmland to help protect farmland and contribute to the development of viable farming activities (Bryant 2011). This municipality has a large land base and small population, at 920 residents in 2011. Farming in this area is primarily organic, with a strong orientation to local marketing and food baskets. The project involved two of the co-authors of this chapter, Bryant and Chahine (2010).

The project, initiated by Senneville's small group of farmers, aimed to guarantee a long-term future for their agricultural activities. The farmers made the first contact in May 2009 with the research team led by Bryant. Soon the farmers sought to involve other, mostly non-agricultural, actors, to support their project. The researchers mobilized a large number of collective actors.

This led to a colloquium in July 2009 in which about 100 representatives of different collective actors were present (e.g., elected municipal officials, the Agglomeration of Montreal, the Ministry of Agriculture, various community organizations, and the Quebec Farmers Union) as well as some landowners. A collective vision for the territory was constructed; although agricultural activities remained at the core of this vision, it integrated other functions of farmland and the surrounding territory, such as conservation and leisure activities. Meetings were planned and carried out, leading in the spring of 2011 to presentations to and formal interventions in the municipality about the land-use plan. The project is ongoing and has become integrated into the emerging Montreal Greenbelt.

In all of these examples, one point emerges clearly: the importance of a continuous management process involving municipal, agricultural, and governmental representatives and, of course, farmers themselves. It is also important to recognize how many of the peri-urban agricultural spaces are managed in effect by a small number of farmers and their families. This makes it all the more important to ensure that an intensive management approach is adopted, with direct involvement of and constant dialogue with the farming population. With so few farming families involved in many of these spaces (e.g., Longueuil, St.Hilaire, and Senneville), not taking into account the circumstances of each farm family could ensure the demise of the agricultural zones concerned.

CONCLUSION

This brief overview of the farmland protection program in Quebec illustrates two points:

1. It is important to provide a broad (provincial) framework that takes account of the essentials of farmland protection, notably the quality of farmland resources, and the general considerations needed to maintain and develop a viable agricultural sector.

2. It is equally important to ensure the maintenance and development of a socially and economically productive agriculture by espousing a flexible approach to the management of each peri-urban agricultural space, taking into account local characteristics, emerging alternative forms of "agricultural production," and farmers and their families.

Thus, the broader framework for farmland protection should permit the integration of a provincial-level perspective with needs, creativity, and flexibility that can only come from local initiatives. This holds out hope for better integration of land-use planning for peri-urban agricultural spaces with development planning for agricultural production and other functions of agricultural land.

The result, well on its way to fruition in several parts of Montreal's peri-urban zone, is a combination of farmland protection and development that integrates the many actors who have an interest in, and influence on, agricultural production and other functions of farmland in such areas, including farmers, their representatives, municipal councils and planning departments, urban consumer groups, and environmental interest groups. This has been encouraged and supported by the Ministry of Agriculture through its program of supporting the PDZA in several MRCs in the CMM over the past four years. Finally, other initiatives, such as that on Île Bizard—with its humanized landscape initiative, development of an equitable and sustainable food system for Montreal (CRÉ de Montréal 2011), and general increase in recognition given to urban agriculture in the Montreal area (including traditional "urban agriculture" and "peri-urban agriculture")—have put agricultural planning on the political map and agenda and contributed substantially to reinforcing the complementary links among actors involved in the issues.

REFERENCES

Barreteau, G. 1997. *Le tourisme : Une nouvelle fonction pour l'agriculture périurbaine, l'exemple de Mont-Saint-Hilaire.* Mémoire de fin d'étude, École Nationale Supérieure Agronomique de Montpellier, Montpellier.

Bryant, C.R. 1984. "The Recent Evolution of Farming Landscapes in Urban-Centred Regions." *Landscape Planning* 11: 307–26.

———. 1995a. "The Role of Local Actors in Transforming the Urban Fringe." *Journal of Rural Studies* 11: 255–67.

———. 1995b. "Interests, Interest Groups, and the Rural Environment and the Challenge of Modelled or Perceived Futures for the Rural Environment." In *Scenario Studies for the Rural Environment* 25–33. Dordrecht: Kluwer Academic Publishers.

———. 2011. "Les dynamiques des agricultures périurbaines autour de Montréal : Défis et opportunités au service de la société métropolitaine." In *Panorama des régions du Québec, édition 2011* 13–28. Québec: Institut de la Statistique du Québec. http://www.stat.gouv.qc.ca/publications/regions/panorama.htm.

———. 2012a. "Transformations agricoles de la couronne périurbaine montréalaise et développement durable agricole." Invited presentation, Forum métropolitain sur le mise en valeur de la zone et des activités agricoles, Communauté Métropolitain de Montréal, le 1 novembre 2012.

———. 2012b. "The Social Transformation of Agriculture: The Case of Québec." In *Social Transformation in Rural Canada: New Insights into Community, Cultures, and Collective Action*, edited by J.R. Parkins and M.G. Reed, 294–310. Vancouver: UBC Press.

Bryant, C.R., and G. Chahine. 2010. "Pour un rapprochement entre urbanité et agriculture, ou la protection de l'agriculture par le développement local et la multifonctionnalité." *Villes et campagnes, une complicité à cultiver*, special issue of *Développement social* 11 (2): 36–37. http://www.revueds.ca/sommaire-volume-11-no-2.aspx.

Bryant, C.R., S. Desroches, and P. Juneau. 1998. "Community Mobilisation and Power Structures: Potentially Contradictory Forces for Sustainable Rural Development." In *Dimensions of Sustainable Rural Systems*, Netherlands Geographical Studies No. 244, edited by I.R. Bowler, C.R. Bryant, and P.P.P. Huigen, 233–44. Utrecht: Groningen.

Bryant, C.R., and T.R.R. Johnston. 1992. *Agriculture in the City's Countryside.* London: Pinter Press; Toronto: University of Toronto Press.

Bryant, C.R., and L.H. Russwurm. 1979. "North American Farmland Protection Strategies in Retrospect." *GeoJournal* 6 (6): 501–11.

Bryant, C.R., L.H. Russwurm, and A.G. McLellan. 1982. *The City's Countryside: Land and Its Management in the Rural-Urban Fringe.* London: Longman.

Bunce, M. 2008. "Les défis de la gouvernance du Plan de la Ceinture Verte de la Greater Golden Horseshoe, Ontario, Canada." In *Territoires périurbains et gouvernance: Perspectives de recherche*, edited by S. Loudiyi, C.R. Bryant, and L. Laurens, 75–82. Montréal: Laboratoire de Développement Durable et Dynamique Territoriale, Université de Montréal.

Charbonneau, K. 2010. "Continuum ville/campagne: Un projet territorial intégré." *Villes et campagnes, une complicité à cultiver*, special issue of *Développement social* 11 (2): 24. http://www.revueds.ca/sommaire-volume-11-no-2.aspx.

City of Laval. 2012. "Laval Technopôle." http://www.lavaltechnopole.com/the-poles/agropole-en.html.

CMM (Communauté Métropolitaine de Montréal). 2012a. http://www.cmm.qc.ca.

——. 2012b. "PMAD : Plan Métropolitain d'Aménagement et de Développement." www.pmad.ca.

CPTAQ (Commission de Protection du Territoire Agricole du Québec). 2007 (updated 2011). http://www.cptaq.gouv.qc.ca.

CRÉ (Conférence Régionale des Élus) de Montréal. 2011. *Plan de développement d'un système alimentaire équitable et durable de la collectivité montréalaise*. Montréal: CRÉ de Montréal. http://credemontreal.qc.ca/cre_evenements/plan-de-developpement-dun-systeme-alimentaire-durable-et-equitable-de-la-collectivite-montrealaise.

Darly, S. 2001. "La multifonctionnalité du territoire agricole : Enjeux et réalités pour la ville de Laval (Québec)." Diplôme d'Ingénieur Agronome thesis, Institut National Agronomique Paris-Grignon.

Deslauriers, P. 1995. "Zonage agricole et utilisation du sol en milieu rural : Quelques cas québécois." In *Actes, premier colloque du Groupe d'Étude de l'IGU sur Le Développement Durable de Systèmes Ruraux* 168–78. Montréal: Département de Géographie, Université de Montréal.

Dumoulin, O., and C. Marois. 2003. "*L'émergence des stratégies de développement des espaces agricoles périurbains* : Le cas des municipalités de banlieue de la région métropolitaine de Montréal." *Canadian Journal of Regional Science* 26 (2–3): 337–58. http://www.cjrs-rcsr.org/archives/26-2-3/8-Dumolin-Marois.pdf.

Frej, S., M. Doyon, D. Granjon, and C.R. Bryant. 2003. "La construction sociale des localités par des acteurs locaux: Conceptualisation et bases théoriques des outils de développement socio-économique." *Interventions économiques* 30: 1–. www.teluq.uquebec.ca/interventionseconomiques.

Furuseth, O.J., and M.B. Lapping, eds. 2001. *Contested Countryside: The Rural Urban Fringe in North America*. Perspectives on Rural Policy and Planning. Aldershot, UK: Ashgate Publications.

Giddens, A. 1984. *The Constitution of Society*. London: Polity Press.

Gouvernement du Québec. 1994. *Les orientations du gouvernement en matière d'aménagement du territoire*. Québec: Gouvernement du Québec.

——. 1997. *Les orientations du gouvernement en matière d'aménagement: La protection du territoire et des activités agricoles*. Québec: Gouvernement du Québec.

Granjon, D. 2004. "La multifonctionnalité de l'espace métropolitain et le développement des activités agro-touristiques en zone périurbaine de Montréal." PhD diss., Université de Montréal.

Gumuchian, H., E. Grasset, R. Lajarge, and E. Roux. 2003. *Les acteurs, ces oubliés du territoire*. Paris: Anthropos.

MAPAQ (Ministère de l'Agriculture, des Pêcheries et de l'Alimentation du Québec). 2012a. Plan de développement de la zone agricole (PDZA). Quebec: Quebec Ministry of Agriculture.

———. 2012b. Guide d'élaboration d'un PDZA. Quebec: Quebec Ministry of Agriculture.

Marois, C. 2008. "La survie des espaces agricoles périurbains montréalais: Une gouvernance qui se cherche." In *Territoires périurbains et gouvernances: Perspectives de recherche*, edited by S. Loudiyi, C.R. Bryant, and L. Laurens, 57–64. Montréal: Laboratoire de Développement Durable et Dynamique Territoriale, Université de Montréal.

———. 2010. "Pour une meilleure compréhension de la Loi sur la protection du territoire et des activités agricoles." *Villes et campagnes, une complicité à cultiver*, special issue of *Développement social* 11 (2): 12–13. http://www.revueds.ca/sommaire-volume-11-no-2.aspx.

Marsden, T., J. Murdoch, P. Lowe, R. Munton, and A. Flynn. 1993. *Constructing the Countryside*. Boulder, CO: Westview Press.

Montminy, D. 2010. "Les impacts de l'étalement urbain sur les zones agricoles dans la région métropolitaine de recensement (RMR) de Montréal." MA thesis, Université de Montréal.

Ontario. Ministry of Municipal Affairs and Housing. 2004. *Toward a Golden Horseshoe Greenbelt: Greenbelt Task Force Advice and Recommendations to the Minister of Municipal Affairs and Housing*. Toronto: Ontario Ministry of Municipal Affairs.

Planchenault, M. 2001. *Planification particulière du territoire agricole de la MRC/Ville de Longueuil : Diagnostique, projet pilote continuum ville-campagne*. Ville de Longueuil: MRC de Champlain.

———. 2008. "Le projet pilote 'Continuum ville-campagne' de Longueuil: De la résilience territoriale à la gouvernance responsable." In *Territoires périurbains et gouvernances: Perspectives de recherche*, edited by S. Loudiyi, C.R. Bryant, and L. Laurens, 65–74. Montréal: Laboratoire de Développement Durable et Dynamique Territoriale, Université de Montréal.

RMM (Région Métropolitaine de Montréal). 2001. *Cadre d'aménagement et orientations gouvernementales pour la Région Métropolitaine de Montréal: Une vision d'action commune, Région Métropolitaine de Montréal 2001–2021*. Montréal: Région Métropolitaine de Montréal.

Tanguay, L., and B. Arpin. 2001. *L'agriculture dans la Communauté Métropolitaine de Montréal, portrait sommaire*. Québec: Ministère de l'Agriculture, de l'Alimentation, et de la Pêche.

Thibodeau, J.C. 1984. "Une urbanisation mieux contenue, une agriculture qui se régénère." *Cahiers de l'Institut d'Aménagement et d'Urbanisme de la Région d'Île-de-France* 73: 26–39.

Thibodeau, J.C., M. Gaudreau, et J. Bergeron. 1986. *Le zonage agricole, un bilan positif*. Montréal: INRS-Urbanisation.

UPA (l'Union des Producteurs Agricoles). 2003. *Projet d'énoncé de vision stratégique du développement économique, social, et environnemental de la CMM*. Mémoire de l'Union des Producteurs Agricoles présenté à la Communauté Métropolitaine de Montréal.

CHAPTER THREE

Farmland Preservation in Ontario

WAYNE CALDWELL, STEW HILTS, AND BRONWYNNE WILTON

Diary entry, 1 June 1876[1]

As I look out over my lush green fields I'm amazed at what we've accomplished over the last 30 years. When my wife and I first arrived here we were confronted with forest as far as the eye could see. There was a track through the bush that served as a road that led to the beginnings of a town a couple of miles away. I had many other neighbours who had just immigrated to this wilderness who, like me, were beginning to clear the land for farming. The potential of this land for agriculture was immediately obvious. Soon we were growing wheat, oats, peas, potatoes, turnips and hay. We even had an apple orchard. We also had livestock—milk cows, horses, sheep and pigs. Our strength was our diversity.

Since those early days we had built ourselves a fine frame house. We have also built a community—we've got a school and church and better roads and many neighbours. We work together and with the town up the road we've got all that we need right here.

I'm a lucky man. We've raised five children on this farm and my youngest boy will take over this farm next year. He'll have his two brothers with their families across the road and other neighbours to work with.

Maybe this farm will be in the family for generations to come.

Diary entry, five generations later, 1 June 2013

It's been a good day. I finally got the crop in—70 acres of corn, 90 acres of soy beans, 50 acres of wheat, and 70 acres of alfalfa. It's a good thing that the hay will be a week or two yet—it'll give me a chance to catch up.

This farm where we live has been in the family for five generations now. I took it over from my father twelve years ago. I think I have the best farmland in Canada. I don't know about the future, though. It seems like I'll soon be surrounded by the city. I'm not sure what to do. What with the taxes going up, people cursing me as I drive down the road with my tractor, trespassers, and complaints about the odours from the dairy cows, it's tempting to pack it in and move forty miles to the west. Last year I thought about expanding, but I couldn't even get a building permit because of the proximity to the residential lots across the road. At least I can get a good price for the farm.

I just don't see how this farm fits in. I also don't see my kids taking over the farm—too much expense, too many hassles, and all the time I sit on a farm that I can easily sell allowing me to relocate elsewhere.

I don't know, I think I'll talk to my wife, and maybe, just maybe, I'll give that real estate agent a call, you know, the one who dropped in last summer. . . .

And so the process of conversion from rural land to urban land continues.

Quality farmland used for agriculture for generations is being removed from production and converted to residential, commercial, industrial, recreational, and institutional uses in support of a growing urban population. The economics of urban growth, and the difficulties of farming combined with the specific challenges for agriculture in the urban shadow, contribute to the loss of farmland. To a certain degree, these processes, where there is significant urban growth, are unavoidable. As millions of people are added to Ontario's population, there will be impacts. At a minimum, however, it is reasonable to expect that agricultural land will be valued as an important resource and that public policy will ensure a predictable pattern of land conversion, thereby preventing the premature and unnecessary urbanization of farmland. It is also reasonable to expect that public policy will protect the most unique agricultural resources in the province. Finally, substantial non-farm development in rural areas can significantly diminish the ability of farmers to respond to changing economic, climatic, and production circumstances, resulting in a corresponding reduction of flexibility in agricultural production. Fortunately, through appropriate public policy, these issues can be addressed and their damages mitigated.

The public interest as it relates to agricultural land is an underlying principle of this chapter. The public interest reflects a societal view of the common good. It can protect both individuals and broader societal interests. For example, a prohibition against building in a floodplain reflects the public interest of protecting life and property—both at an individual level and at a societal level. It involves both a present and a future perspective. For example, the protection of gravel deposits from having a subdivision built on them is intended to protect the resource for current and future generations. Protection of the public interest also means that at times individual interests, rights, or desires are secondary. For example, an individual might want to clear and drain a wetland for agriculture, but because of the broader societal or public interest related to protection of the wetland the landowner is forbidden from doing so. This is a fundamental concept and underpins both provincial and local planning policy (Leung 2003).

This chapter reviews the basis for agricultural land preservation, reviews existing policies in Ontario, identifies options for the future, and considers evolving policy directions.

THE PROTECTION OF AGRICULTURAL LAND AND THE PUBLIC INTEREST

Policy to protect farmland is based upon the premise that it is in the public interest to protect farmland, farmers, and the farm economy. In the Canadian context generally, and in Ontario specifically, high-capability farmland is a finite resource. Ontario contains 52 percent of Canada's Class 1 land, but only 6.8 percent of the total land area of Ontario is suitable for agriculture, with even less than that remaining accessible to agriculture. Even more remarkable is the excellent climate in Ontario for agricultural production. With an abundance of heat units and rainfall, crops can be produced in Ontario that are not feasibly produced elsewhere in Canada. The result is the highest number of farms in Canada and a diverse and active agricultural industry. And in

2011 Ontario led all Canadian provinces in gross farm receipts with diverse livestock and crops.

Although the importance of Ontario's agricultural industry is generally recognized, the commitment of the public and different levels of government to its long-term protection has wavered. The loss of specialty croplands to urbanization was documented as early as the 1950s, but it was not until 1978 that the province adopted the Foodland Guidelines as policy for planning for agriculture (Krueger 1959). By this time, a number of counties and regions had already developed their own policies. Some, such as Huron and Waterloo, had established stringent policies to protect farmland, whereas others continued to grapple with the issue. Its importance was not lost on the farm community. In 1976, David Brown of the Federation of Agriculture, speaking at the conference Crisis in the Countryside, identified the following objectives for rural land-use policy: "to preserve and encourage a profitable agricultural industry in Ontario, to preserve as much high class agricultural land as possible within the province and to allocate it to agriculture for the continued production of food and fibre; to ensure an adequate supply of high quality food . . . now and in the future without an undue reliance on imports; . . . to assure that agriculture continues to be located on productive land that can be operated as viable and progressive farms and as profitable business; to protect farmers from harassment and complaint from non-farm encroachment into agricultural areas."

In 1994, the Foodland Guidelines were replaced by the Comprehensive Provincial Policy Statement, itself replaced in 1997 by the Provincial Policy Statement. In 2005, the provincial government took a number of actions, including the adoption of a revised Provincial Policy Statement (PPS) with corresponding amendments to the Planning Act and reform of the Ontario Municipal Board, all of which have implications for farmland preservation. The Provincial Policy Statement has been under review since 2010 and has proposed "permitting additional uses on farms and providing flexibility for agricultural-related uses; requiring agricultural areas to be designated in municipal official plans and impacts of non-farm development surrounding agricultural operations to be mitigated; [and] strengthening requirements for the rehabilitation of specialty crop areas that are subject to aggregate extraction" (Ontario, Ministry of Municipal Affairs and Housing 2012). In December 2003, the government released Bill 27, the Greenbelt Protection Act, with the goal of establishing permanent protection of a greenbelt, including a combination of agricultural and rural natural environment lands across the Golden Horseshoe region. The Greenbelt Protection Act was adopted in 2005 amid contention and is undergoing a coordinated review which started in 2015. The Government of Ontario is reviewing the four provincial land-use plans and has proposed changes based on stakeholder and public feedback to improve the plans and to better strategize for the future of the Golden Horseshoe region. The review includes discussions on the possible expansion of the greenbelt.

Although evolving policy in Ontario suggests that there is renewed interest in farmland preservation, there are many layers to the debate. Some argue that, because of low commodity prices, agricultural surpluses, inexpensive food imports, and overall pessimism in certain agricultural sectors, there is no point in preserving farmland. Conversely, others argue that the protection of farmland should be a priority, both nationally and locally, based upon the need to protect the potential for food production, enhance the role of agriculture in local and national economies, and deal with climate change and rising energy costs.

There are five overriding reasons that the protection of farmland is in the public interest:

1. *Food production.* Food is an essential human need; life depends on it. At a national level, the ability of a country to feed its citizens is important, and at a local level access to a variety of fresh food options is vital to healthy communities. Climate change, through more frequent and severe storms along with record-breaking droughts, is changing agricultural production and driving home the need for a land base able to adapt to changing conditions to continue to produce food.

2. *Food security.* In the past, food security would have been thought of exclusively as the ability of a nation to feed itself, but in a post-9/11 world, with threats of related international terrorism, food security takes on a new meaning. In addition, the production of food in Canada occurs under rules and regulations set by Canadians. The use of pesticides, and the production and processing of food in both this country and abroad, are under the scrutiny of the public, media, and government.

3. *Economic contributions.* Agriculture and related food industries provide livelihoods for millions of Canadians.

This is important nationally but even more so in many rural economies, in which the agri-food sector is often the economic backbone of a community.

4. *Stewardship and amenity of the countryside.* The countryside is a resource for all Canadians. Many people take pride and enjoyment in driving through a rural landscape with a patchwork of fields and crops and attractive farm landscapes. Even more important is the role that farmers play in producing quality food while maintaining water quality, clean air, and much of southern Ontario's natural heritage. As climate change mitigation is discussed more frequently, conservation authorities and governments are realizing the importance of farmland as wildlife habitat and carbon sequestration areas.

5. *A resource for future generations.* The protection of farmland is for today and tomorrow. Future generations will benefit from wise decisions made today that protect farmland for a growing population.

There are also a number of specific reasons to limit rural non-farm development. Although these reasons are reviewed in more detail in Chapter 8, by Caldwell, Churchyard, and Dodds, it is important to identify the impacts of such development as justification for appropriate land-use planning. Rural non-farm development can

- fragment the land base and consume farmland;
- be a potential source of conflict with agriculture;
- fundamentally change the rural community;
- lead to the introduction of restrictions on farmland;
- increase the cost of providing services for the municipality;
- have environmental and servicing impacts; and
- detract from the rural aesthetic.

In Ontario, farmland preservation policies have been debated for years. From a certain perspective, there has been reasonable support, but success has varied by region and municipality. This pattern reflects the policy-led system that balances provincial and municipal interests. Details of this system are presented in the next section.

EXISTING FARMLAND PRESERVATION POLICY AND APPROACHES IN ONTARIO

The Planning Act provides the framework for farmland preservation policy in Ontario. It enables municipal land-use planning and establishes the basis for provincial policy.

The implementation of provincial policy reflects the relationship between municipalities and the province. This relationship, based upon a legal and jurisdictional framework, establishes the basis for municipal planning. Municipal corporations are not autonomous bodies but limited in their powers by the terms of provincial legislation. The Planning Act, for example, assigns certain authority to municipalities while retaining an overriding provincial role. This process requires that due consideration is given to provincial interests (which the Planning Act defines as including the protection of agricultural resources) and provincially adopted policy statements. The PPS provides policy direction on matters of provincial interest related to land-use planning. The PPS is intended to promote a policy-led system to ensure the long-term wise use and protection of the province's resources.

From the provincial viewpoint, the PPS (revised in 2014) has further promoted an approach to planning that emphasizes the importance and priority of agriculture, farm development, and rural areas. The PPS guides the development of official local plans and affects the review of individual developments. Municipal councils, provincial ministries, commissions, and the Ontario Municipal Board are required to "be consistent with"[2] the PPS. In many respects, the agricultural policies included in the current PPS maintain the thrust of the original Foodland Guidelines. The provincial interest in agriculture as stated in the Planning Act is "the protection of the agricultural resources of the Province" (Government of Ontario 1990). Key aspects of the Provincial Policy Statement (2005) that relate to agriculture are summarized below:

1. Prime agricultural areas will be protected for agriculture. Prime agricultural lands include specialty croplands and Classes 1–3 soils. These areas are restricted to agricultural uses (e.g., crops and livestock), secondary uses (e.g., home occupations), and agriculture-related uses (e.g., grain-drying facilities).

2. Urban areas will be expanded only where existing designated areas in the municipality do not have sufficient land supply to accommodate projected growth, and expansions into prime agricultural areas are permitted only where there are no reasonable alternatives.

3. Limited non-residential uses are permitted on prime agricultural land provided that there is a demonstrated need and no reasonable alternative location.

4. Lot creation in prime agricultural areas is generally discouraged and will be permitted for agricultural uses, related uses, and surplus residences.

5. Extraction of mineral resources is permitted on prime agricultural lands, including the potential for extraction below the water table.

6. New development and new or expanding livestock facilities will comply with the minimum distance separation formulae.

7. In prime agricultural areas, agricultural uses and normal farm practices will be promoted and protected.

In some respects, it is difficult to evaluate provincial initiatives to protect farmland. Past policies have had effects, but at a provincial scale the conversion of farmland into non-farm uses has continued. Although some would argue that this is inevitable, perhaps more telling are a number of examples from across the province, where scattered residential development has continued relatively unabated in some municipalities (Caldwell and Dodds-Weir 2003). Although relatively high densities have been achieved in a number of Canadian cities, the conversion of key agricultural lands to urban uses continues, with 4.8 percent of remaining farmland in Ontario lost between 2006 and 2011 (Condon 2004; Kulasekera 2012). Confusing the issue further are some highly successful local policies (Caldwell and Dodds-Weir 2003). However, with the implementation of the original Places to Grow Act in 2005 and the related Growth Plans released for the Greater Golden Horseshoe (2006), northern Ontario (2011), and the Simcoe sub-area (2012), this growth is focused on specific growth centres. These growth targets have raised contentious issues with local and regional governments, but they represent a desire to curb sprawl and therefore aid in farmland preservation. For example, the Greater Golden Horseshoe Growth Plan requires that, by 2015, 40 percent of new residential development be built within existing urban limits, reducing the amount of sprawl that characterized development of the late 1990s and early 2000s (Ontario, Ministry of Infrastructure 2012).

Although the PPS has had successes, it has also been criticized. Revisions adopted by the Ontario government in 2005 have strengthened policy on the protection of agricultural land, but several issues still haven't been addressed. Some of the identified weaknesses of the PPS include that it is discretionary and not absolute (i.e., not legislation), successful farmland preservation has been dependent on the cooperation of local municipalities, policy has often been implemented in isolation of other issues in the rural community, and finally the overall commitment of the province to farmland preservation has been questioned. Where the Provincial Policy Statement has been most successful, it has been accompanied by locally developed and supported municipal policy often more holistic, comprehensive, and restrictive than provincial policy.

Although decisions on the best approach to protecting farmland in Ontario are inherently political, researchers and planners can play roles in helping to suggest options and best practices for farmland preservation. The following section builds upon recognition that farmland preservation is in the public interest and that improvements to the existing system in Ontario are required.

POLICY DIRECTIONS FOR ONTARIO

Ontario pursued a number of new directions from 2004 to 2006. Since then the province has been focused on implementing the Greenbelt Act and Places to Grow Act and oncompleting the five-year review process for the PPS, and it has not undertaken any new initiatives. The changes that occurred during the mid-2000s in the approach to farmland preservation can be placed into three categories:

- *Building on the existing system.* The existing system has a number of attributes and in some parts of the province has been very successful in ensuring that the protection of farmland is given a high priority in planning decisions. This includes some "minor tinkering," such as the 2005 revisions and the 2010 revisions to the PPS, that can have significant and positive results.

- *Moderate changes to the existing system.* New policies and programs for Ontario generally do not require any changes to the existing legislative framework and build upon voluntary and incentive-based approaches. There is the potential to achieve a high level of public support since many of these approaches are not regulatory. The Ontario Farmland Trust, created in 2004, is an example of a more moderate change.

- *A radical departure from the existing system.* Provinces and states across North America have taken bold steps to establish new tools and methods of farmland preservation. Many of these tools might require new legislative frameworks: they might include enhanced regulation or come with a substantial budgetary requirement. The Greenbelt Act, adopted in 2005, is an example of a more radical departure and it was reviewed in 2015 along with the Oak Ridges Moraine Conservation Plan and the Niagara Escarpment Plan. In 2016, the provincial government was still considering potential revisions to these plans.

BUILDING UPON THE EXISTING SYSTEM

Four key options, which represent relatively minor changes and build upon the existing system, are identified.

1. *Amendment of the Planning Act.* In 2005, the Ontario government amended the Planning Act, requiring that municipal decisions be "consistent with" the PPS. This amendment replaced the words *have regard to*. This change directs municipalities and the Ontario Municipal Board to place a stronger emphasis on compliance with the agricultural policies included in the PPS.

2. *Public and political awareness.* Support for farmland preservation is likely to benefit from an informed public. Ball and colleagues (2002) noted the extent to which residents of Pennsylvania were engaged by the governor's office in discussions on preserving farmland. This involvement appeared to significantly raise the profile of the issue and generated public and political support.

3. *Approvals for farmland conversion with the province.* During the mid-1990s, the provincial government, in an attempt

to streamline the planning process, delegated a number of planning approvals to municipalities (sometimes putting the province in the position of having to appeal local planning decisions). Although this approach came with certain advantages, it removed the provincial policy review, leaving approvals vulnerable to local political influences.

4. *Amendments to provincial policy.* The revised PPS (2005) eliminated the following categories of severance: retirement and residential infilling in rural areas. The province also moved to permanently protect specialty cropland. Increased justification for conversion of farmland to urban uses was also adopted.

MODERATE CHANGES TO THE EXISTING SYSTEM

A number of suggestions are useful in planning to protect agricultural land. These options are identified as moderate because they represent a new direction for the province or involve a more substantive change.

1. *Incentivize good municipal practices.* A number of municipalities in Ontario have had remarkable success protecting agricultural land (Caldwell, Dodds-Weir, and Thomson 2003). Perth and Huron Counties, for example, have excellent records of limiting rural non-farm development. Waterloo Region has some of the highest growth rates in Canada, yet it has managed to virtually eliminate rural non-farm development. Although there has been significant growth in population, it has generally occurred in relatively dense, planned expansions of urban centres. Grey County at the start of the 1990s had such a poor record that the province stepped in and suspended its authority to approve severances. By the end of the decade, however, Grey County had developed a new official plan, public support had been mobilized, and rural severances had significantly decreased. Although many of these areas have much more stringent policies than provincial policy, there is no guarantee that they will maintain their high standards. To incentivize good practices, a number of actions can be taken, ranging from financial assistance to the planning system to planning awards bestowed by the province. Some

form of incentive or special recognition can lead to ongoing
and enhanced commitment to responsible policies.

2. *Build upon the concept of Smart Growth.* This concept has
received much attention (Daniels 2001). In some ways, it
captures what planners have advocated for decades: good
planning decisions yield multiple benefits, including reduced
costs for infrastructure, an enhanced assessment base,
protection of natural areas, quality urban environments, and
protection of farmland. Smart Growth also holds promise
for rural communities to benefit from development without
risking the agricultural resource. This concept has been built
into a number of actions, including the Places to Grow Act and
related Growth Plans, but it can continue to be a powerful tool
in helping to convince the public (and local politicians) of the
wisdom of protecting agricultural land.

3. *Adopt special policies for special areas.* Ontario is blessed with
a number of unique agricultural resources. For example, the
evolving estate wine industry in the Niagara Region can offer a
unique agri-tourist resource for generations to come. Although
enhanced protection of this area is required, certain support
activities not normally allowed in an agricultural area might
be appropriate. In wine regions around the world, it is not
unusual to find wine production and related tourist facilities
close to each other.

The province also has a wide range of agricultural practices and community interests, and interest in farmland preservation varies across the province. This has been addressed to some extent through specific area plans, such as the Niagara Escarpment Plan (2005), the Oak Ridges Moraine Conservation Plan (2002), and the Lake Simcoe Protection Plan (2009), yet the opportunity remains to resolve other regional issues through this avenue.

1. *Land trusts and agricultural easements.* Land trusts can
permanently protect farmland. Through the acquisition of
lands or easements,[3] the trust can permanently maintain land
for agricultural production. The Ontario Farmland Trust,
incorporated in 2004, has a mandate to focus on farmland
preservation through the use of easements, research, and

education to improve farmland protection policies (Caldwell, Hilts, and Thomson 2004). Although the total acreage protected using this mechanism is likely to be limited, the enhanced profile of protecting farmland, particularly in the urban shadow, is likely to have benefits for generations to come. Moreover, this initiative can be pursued by community members without the need for government assistance.

2. *Provincial development plan.* Over the years, Ontario has proposed a number of initiatives to direct development (e.g., the 1971 Design for Development). Results have been mixed, but there is a clear recognition that urban expansion does not need to occur on the province's best farmland. For example, growth north and east of Toronto is likely to have less impact on agriculture than growth south and west of it. An opportunity exists to develop realistic assessments and plans that guide development toward certain areas of the province. The Places to Grow legislation and related Growth Plan for the Greater Golden Horseshoe are important for this area.

3. *Public investment in infrastructure.* This option can assist at both the provincial level and the local level. This tool appears to have been used substantially in the United States to guide development (Caldwell et al. 2004). At the provincial level, decisions related to highways, public transit, and water and sewage facilities can help to direct growth toward one area or away from another. Likewise, municipalities can use it as a tool to achieve higher densities, avoid sprawl, and encourage infilling and direct growth.

A RADICAL DEPARTURE FOR ONTARIO

A new legislative framework can launch innovative, bold, and at times challenging new directions. Previous examples in Ontario include the Oak Ridges Moraine Act (2001) and the Greenbelt Act (2005). Although both legislative initiatives partially address the preservation of farmland within defined geographic areas, there is the opportunity to establish new options for protecting farmland across the province.

There are examples in other provinces of what would be a radical departure for Ontario—notably the BC Agricultural Land Commission Act

(1973) and the Quebec Loi sur la protection du territoire agricole (1978). Both involve powerful provincial commissions (Bryant 1994). There are also examples in the United States, where the Purchase and Transfer of Development Rights is widely used as a mechanism to guide development (Daniels and Bowers 1997). Other American programs often focus on financial incentives and include Live Near Your Work programs, priority-funded areas, and job creation tax credits (Caldwell et al. 2004).

Although this chapter does not come to a conclusive recommendation concerning the most appropriate direction for Ontario, it does suggest some key criteria to evaluate these options, as described below.

CRITERIA TO EVALUATE POLICY

For a policy or new program to be adopted, it needs to provide significant benefits and have a high probability of successful implementation. In evaluating a new policy, politicians and policy makers consider a number of criteria to determine its appropriateness. In terms of farmland preservation, the following list provides key tests for any new action.

1. Any new action needs to lead to improved farmland preservation and have a high probability of acceptance. For example, the best policy for farmland preservation might be to issue a prohibition on urban expansion, but the difficulty is that, with 4.4 million more people expected in Ontario between 2011 and 2036, this would be an unworkable solution. Likewise, legal issues, property rights, and public perceptions need to be considered.

2. Public, political, and farm support will be essential. In the rural community, issues of farmland preservation can be emotionally charged, for people are concerned about land values and want to preserve their perceived property rights. Where municipalities have successfully limited non-farm development in the past, it appears to have been accompanied by a local belief that the protection of farmland is in the best interests of individual farmers (Caldwell and Dodds-Weir 2003). In fact, the agricultural value of property in some parts of the province has increased because of the absence of non-farm development. Support from the public can lead

to political support for any public expenditure to protect farmland.

3. There are also a number of legal issues related to protecting farmland, both in terms of actions that might be taken (the legality of various options) and in terms of respecting that farmland is privately owned.

4. Options for protecting farmland also need to be financially sound. In Pennsylvania, for example, $400 million has been spent to purchase development rights to protect 200,000 acres of farmland (Ball et al. 2002). A similar program in Ontario might be desirable, but securing this amount of money would present obvious obstacles.

5. Across Ontario, there are diverse attitudes toward agriculture and the preservation of farmland. Historically, provincial policy has accommodated the interests of the entire province. The agricultural industry, however, varies across the province, and so does support for farmland preservation. In southwestern Ontario, for example, some municipalities have not allowed a rural non-farm severance on prime farmland in decades. The challenge is to find a policy acceptable across the province or acknowledge the basis for some regional variation.

6. The role of the province versus the role of the municipality is also a key consideration. The most success in Ontario has occurred at a municipal level, where a given municipality, based upon extensive public support, has developed and implemented policy more supportive of farmland preservation than that established by provincial policy. Would a province-wide system such as the BC Agricultural Land Reserve enhance this strong local support or lead to a weaker policy in key areas of the province?

7. Serious initiatives aimed at farmland preservation also need to accommodate a long-term perspective. Agricultural investment will most likely occur where there is certainty that farmland will remain in production for many years. Farmland preservation strategies need to offer this certainty over many years.

SUMMARY

Many challenges confront initiatives to preserve agricultural lands. In North America, there continues to be a challenge of abundance—high levels of food production contributing to low commodity prices. Related to this challenge are differing philosophical views on the role of government in protecting farmland as a public interest. This is further complicated by financial and legal issues and the fact that farmland is privately owned by thousands of individual farmers and landowners who have diverse aspirations for their properties.

Despite the challenges that face initiatives to protect farmland, one undeniable fact remains: decisions made today will fundamentally affect options available to future generations. The ability to produce food, to regulate the system of production to reflect the values of society, to maintain the important economic contributions of agriculture, and to retain the important role that farmers play in managing the countryside is dependent on retaining farmers and the lands essential to their livelihood.

This chapter began with comparative perspectives from 1876 and 2013. If we look ahead to 2036, the next generation will look back at the decisions made today and pass judgment on the success of the current generation in planning for agriculture. In 2036, Ontario will have a significantly larger population. Land will be converted from agricultural uses to urban uses, and there will be ongoing changes in agriculture and rural communities. But will the best farmland be available for farming? Will farmers have a predictable land-use system in which they can continue to invest with the knowledge that farming is the priority use in a given area? Will agriculture thrive in proximity to urban centres as well as in more rural parts of the province? Will agriculture remain a significant part of Ontario's economy? The answers to these questions rest in the options brought forth in this chapter and in the political will to take appropriate action. Important decisions made today by municipal and provincial politicians, planners, the public, and farmers will determine the legacy left to future generations.

NOTES

1 This entry is inspired by *The Settlement of Huron County* by James Scott (1966). Both this and the entry that follows are dramatizations by the co-authors of this chapter.

2 Different provincial governments have vacillated between the words *have regard to* and *be consistent with*. The latter phrase is viewed as having a stronger emphasis on provincial policy.

3 Easements do not necessarily need to be held by a land trust. They can be held by governments, other NGOs, or even individuals.

REFERENCES

Ball, J., E. Brockie, W. Caldwell, J. Marks, M. Nelson, J. Parsons, H. Rudy, S. Stonehouse, S. Weber, C. Weir, and M. Williams. 2002. "Reflections on Planning: Pennsylvania vs. Ontario." *Plan Canada* 42 (2): 30–33.

Brown, D. 1976. "Approaches to Policy." In *Crisis in the Countryside: Proceedings of the Summer and Fall Meetings of the Ontario Chapter, Soil Conservation Society of America, St. Jacobs and Jarvis*, 35–37. Thornhill.

Bryant, C.R. 1994. "Preserving Canada's Agricultural Land." *Plan Canada*: 34: 49–51.

Caldwell, W.J., and C. Dodds-Weir. 2003. "An Assessment of the Impact of Rural Non-Farm Development on the Viability of Ontario's Agricultural Industry." Phase 2 Report. University of Guelph. http://www.waynecaldwell.ca/development.htm.

Caldwell, W., C. Dodds-Weir, R. Dykstra, S. Hilts, and S. Thomson. 2004. "New Strategies for Farmland Preservation: Sizing Up the American Experience." *Ontario Planning Journal* 19 (2): 10.

Caldwell, W., C. Dodds-Weir, and S. Thomson. 2003. "Rural Non-Farm Development and the Future of Ontario Agriculture." *Ontario Planning Journal* 18 (6): 3–4.

Caldwell, W., S. Hilts, and S. Thomson. 2004. "Farmland Preservation and Land Trusts: New Options for Ontario." *Ontario Planning Journal* 19 (1): 8–9.

Condon, P. 2004. *Canadian Cities, American Cities: Our Differences Are the Same*. University of British Columbia. http://www.fundersnetwork.org/usr_doc/Patrick_Condon_Primer.pdf.

Daniels, T. 2001. "Smart Growth: A New American Approach to Regional Planning." *Planning Practice and Research* 3–4: 271–81.

Daniels, T., and D. Bowers. 1997. *Holding Our Ground: Protecting America's Farms and Farmland*. Washington, DC: Island Press.

Government of Ontario. 1990. *Planning Act, RSO 1990, c. P.13*. Government of Ontario. https://www.ontario.ca/laws/statute/90p13.

Krueger, R.R. 1959. "Changing Land-Use Patterns in the Niagara Fruit Belt." *Transactions of the Royal Canadian Institute* 32, Part 2, 67: 39–140.

Kulasekera, K. 2012. *Farm Land Area (Acres) Classified by Use of Land, by County, 2011*. Ontario, Ministry of Agriculture, Food, and Rural Affairs. http://www.omafra.gov.on.ca/english/stats/census/cty32_11.htm.

Leung, H-L. 2003. *Land Use Planning Made Plain*. Toronto: University of Toronto Press.

Ontario. Ministry of Infrastructure. 2012. *Growth Plan for the Greater Golden Horseshoe, 2006: Office Consolidation, January 2012.* Toronto: Province of Ontario.

——. Ministry of Municipal Affairs. 1997. *Provincial Policy Statement.* Toronto: Queen's Printer.

——. Ministry of Municipal Affairs and Housing. 2012. *Provincial Policy Statement Review under the Planning Act: Draft Policies.* Toronto: Province of Ontario.

Scott, J. 1966. *The Settlement of Huron County.* Toronto: Ryerson Press.

Statistics Canada. 2012. *Ontario Provincial Trends.* Statistics Canada. http://www.statcan.gc.ca/pub/95-640-x/2012002/prov/35-eng.htm.

CHAPTER FOUR

The Farmland Preservation Program in British Columbia

BARRY E. SMITH

The Land Commission Act, passed on 18 April 1973, is arguably one of the most important pieces of land-use law passed by the British Columbia legislature. The act, the first of its type in Canada, ushered in a program to preserve the province's limited but often highly valuable farmland resource. The act has also had broader impacts. It has played a role in sustaining the economic and social benefits that have accrued from agriculture to every region of the province. The program has helped to preserve the character of many communities, contributing to their health and livability and helping to maintain valued natural capital. The Agricultural Land Reserve (ALR) has also shaped growth patterns over the past forty years by acting as a de facto urban growth boundary. In so doing, it has contributed to the development of more compact and efficient urban communities and provided an opportunity to address land-use conflicts by ensuring a stable urban-agricultural "edge." The program assists the implementation of Smart Growth principles within urban areas, which in turn contributes to farmland preservation.[1]

As Wilson and Pierce (1982, 17) noted, "it is impossible not to be impressed by the qualities of the political act which grasped the farmland nettle in British Columbia. It is skillful, logical, bold and strong." Creation of the British Columbia Land Commission is considered a major event in the evolution of regional planning in Canada as well as in planning for conservation, resource development, and the environment (Hodge and Robinson 2001). This far-reaching, progressive legislation was a component of "sustainability" long before the term came into vogue, preceding the Brundtland report by fourteen years (Brundtland 1987).

PRECONDITIONS FOR CREATION OF THE AGRICULTURAL LAND RESERVE
A Challenging Geography

After Quebec, British Columbia has the second-largest land area among Canadian provinces. But unlike other provinces, it is characterized by its high mountains and plateaus. Three-quarters of its land base is above 1,000 metres in elevation, and more than 18 percent of the province is rock, ice, or tundra (Cannings and Cannings 1996). This valley-mountain physiography has resulted in a relative scarcity of agricultural land as shown in Table 4.1. Less than 3 percent of the province's land area has an agricultural capability allowing a range of crops (Canada Land Inventory [CLI] Classes 1–4). In 2011, British Columbia accounted for only 4 percent of Canada's land area in farm use. But as an indication of the quality of its agricultural land base and intensity of use, the province accounted for 5.8 percent of Canada's total annual gross farm receipts, 9.6 percent of all farms, 7.7 percent of all dairy cattle, 14.2 percent of all chickens and hens, and 19.4 percent of all land growing fruits, berries, and nuts (Statistics Canada 2011).

Table 4.1. British Columbia's agricultural capability.

LAND CLASS	% OF LAND BASE
LAND CAPABLE OF A RANGE OF CROPS (CLI CLASSES 1–4)	2.70
PRIME AGRICULTURAL LAND (CLI CLASSES 1–3)	1.10
CLASS 1 AGRICULTURAL CAPABILITY	0.06
LAND SUITABLE FOR TREE FRUIT PRODUCTION IN THE ALR	0.04

Source: Smith (1998).

In the narrow valleys that are so agriculturally productive, there has been competition for a variety of settlement uses: urban, industrial, parks and recreation, and infrastructure. The valleys also hold significant habitat and environmentally sensitive areas. As in other jurisdictions, early settlement patterns in British Columbia were often directly associated with areas of prime agricultural land, resulting in urban expansion onto farmland. The province continues to have a highly urbanized population, largely in the southwestern region.[2] As British Columbia experienced rapid growth in the post–Second World War decades, the population nearly doubled between 1951 and 1971 (BC Stats). The province's valley-mountain reality served as a highly visual context for growing public awareness of and concern for the impacts of urbanization on farmland, which would find voice in both regional and provincial land-use policy. As Cannings and Cannings write, "We normally see British Columbia as a series of valleys framed by postcard peaks, but if you stand atop even a small mountain . . . you will always be impressed by the sea of rugged mountains around you and by how small the lush valleys really are" (1996, 155).

Influences, Concerns, and Motivations

A number of preconditions influenced development of the ALR program in British Columbia.

1. The emergence of regional planning sensitized a broad public to the links between urban growth patterns (sprawl) and the implications for farmland preservation.

2. Students in planning schools across the country in the late 1960s and early 1970s were exposed to the concept of regional planning and the importance of resource management and its relationship to urban planning.[3]

3. There was a desire to curtail the loss of prime farmland. Prior to 1973, urbanization was estimated to be converting to non-farm uses between 4,000 and 6,000 hectares of prime agricultural land each year (PALC 1983, 4).

4. There was also a desire to stabilize the agricultural land base to ensure that land improvements designed to enhance agricultural incomes (e.g., dike, drainage, and irrigation systems) were not lost to housing and other non-farm developments (PALC 1973, 8).

5. There was a concern over food security and heavy dependence on external sources of food (PALC 1983, 4–5).

6. There was a desire to reinforce and support local elected government officials concerned with farmland preservation (PALC 1973, 8).

7. It was recognized that many local jurisdictions could not withstand the pressure to change zoning. It was thought that this was a critical point at which almost all land preservation schemes fail (PALC 1973, 7).

8. There was an underlying recognition of global population growth and food shortages.

9. The CLI survey of agricultural capability was fundamental to the designation of the ALR. The CLI provided a biophysical underpinning to the reserve and aided immeasurably its defence.

10. The Social Credit government provided a "stage-setting" piece of legislation in 1971 when it enacted the Environment and Land Use Act. The act, which established the Environment and Land Use Committee (ELUC) of cabinet, was sweeping in its powers and became the tool of choice when the provincial New Democratic Party, in late 1972, "froze" the subdivision and non-farm use of agricultural land (Wilson and Pierce 1982).

The Provincial Election of August 1972

The election, on 30 August 1972, of the New Democratic Party was undoubtedly the most important precondition leading to the development of British Columbia's farmland preservation program. In the lead-up to the election, the NDP was not alone in voicing concern for the preservation of agricultural land. As Baxter (1974) notes, all four major parties outlined an agricultural policy.

The Liberal Party proposed an "Agricultural Land Trust," which would purchase development rights, and the Progressive Conservative Party focused on "long range and systematic planning . . . so [that] the best agricultural land is used for farming and not wasted on other purposes," whereas the Social Credit Party rested on past performance (Baxter 1974, 8). Indeed, in December 1971, the BC Department of Agriculture, with growing concerns over the loss of farmland, prepared a report for Minister of Agriculture Cyril Shelford (see Peterson 1971) recommending that the provincial government undertake a farmland preservation program focused on the Fraser Valley. The proposal recommended the outright purchase of land or the acquisition of development rights. The report did not include an estimate of the program's costs, yet Shelford took the report to cabinet, where Premier Bennett pronounced it "too hot to handle" (Petter 1985, 7).

The NDP's agricultural land-use program was outlined in a booklet, *N.D.P.: A New Deal for People*, and included a proposal to "establish a land-zoning program to set aside areas for agricultural production and to prevent such land being subdivided for industrial and residential areas" (British Columbia New Democratic Party 1972). Despite the pre-election platforms of the Liberals and Conservatives, together the parties gained only seven seats in the election. The NDP alone was in a position to act, and it delivered on its election promise with speed and precision.

THE RAPID PATH TO LEGISLATION

For those interested in the rather chaotic route taken by the NDP government to turn an election promise into legislation in creating the Land Commission Act, Andrew Petter's account is recommended. The process, which appears to have been strongly driven by conviction, was not constructed within a formalized policy mechanism. Petter (1985, 4) asks this question: "How was it possible for a government that lacked a planning structure to put forward so quickly legislation as far-reaching and enduring as the Land Commission Act?"

On 29 November 1972, just two months after the election, Minister of Agriculture David Stupich spoke to a British Columbia Federation of Agriculture (BCFA) convention and stated that "I would not advise anyone to invest in farmland with any intention to develop it for industrial or residential purposes" (quoted in *Vancouver Sun* 1973). As Petter (1985, 9) explains, it was with this speech to this audience that farmland preservation was driven onto the front pages of newspapers across the province, with the result that Stupich effectively "locked" cabinet in on the issue.

British Columbia's farmland preservation program formally began on 21 December 1972 when, after 114 days of being elected, the NDP government enacted Order-in-Council 4483 under the Environment and Land Use Act. Stupich explained that the government had been forced to act quickly after his speech to the BCFA in November because of a rush of rezoning and subdivision applications involving farmland (*Vancouver Sun* 1973).

Order-in-Council 4483 prohibited the further subdivision of land taxed as farmland and lands deemed by ELUC to be suitable for cultivation of agricultural crops. With the passage of Order-in-Council 157 on 18 January 1973, the government further clarified that non-agricultural development was not permitted on any land of two acres or more taxed as farmland, zoned as farmland by a municipality, or having a CLI agricultural classification of Class 1, 2, 3, or 4.

These two orders halted both the subdivision and the non-farm use of agricultural land in British Columbia. These actions became commonly known as the "farmland freeze."

CHOOSING THE RIGHT MODEL: PURCHASE OF DEVELOPMENT RIGHTS OR ZONING?

As legislation was being crafted in late 1972 and early 1973, a debate emerged that would have a lasting effect on the farmland preservation program. This involved a purchase of development rights model versus a land districting (zoning) model.

Petter (1985) describes a serious split in cabinet. The issue revolved around the question of whether development rights were vested in the property owner or the Crown. Stupich had pledged to farmers that the government would compensate them for freezing development on their lands. This was a consistent position of the farm community, and, given the traditional close working relationship between the minister of agriculture and his farm constituents, coupled with his accounting background, his support for this position is not entirely surprising. The earlier farmland preservation proposal of Sigurd Peterson, now Stupich's deputy minister, had also incorporated a plan for the purchase of "development rights." As Petter indicates, the initial draft of the bill reflected the compensation of farmers for their loss of development value (13–17).

Minister of Resources Robert Williams took an opposing view. Influenced by American economist Henry George, he thought that the development value of raw land, like other values, was a public asset not requiring compensation. Although Stupich had not provided a cost estimate of a compensation package, Williams suspected that the cost would be far too high and asked Pearson, his assistant, with a background in land-use planning, to analyze the options. Using Peterson's proposal as a base, Pearson concluded that the purchase of development rights would be so high that it would be more expedient to purchase the land outright (Petter 1985).

In October 1972, just one month before the launch of BC's farmland preservation program, Pearson made his position on the purchase of development rights clear at a workshop—Land Use in the Fraser Valley: Whose Concern?—held in the Lower Mainland. At the workshop, Mayor Douglas Taylor of the District of Matsqui advocated the preservation of farmland by means of the purchase of development rights. Pearson rebuked the proposal. His philosophical departure from the opinion of the mayor was grounded in the need to act in the public or community interest. He commented that the concept of zoning was based upon the viewpoint that development rights were vested not in the landowner but in the Crown and that these rights have subsequently been passed on to municipalities by the province. As a result, a municipality, on behalf of the community, can designate, through zoning, which rights a landowner has to develop the land. Pearson stated that "Mayor Taylor is trying to suggest that these rights are vested in the individual, and that the Queen has to buy them back, but if we were to follow this line of reasoning to its logical conclusion, the Queen would have to, for example, reimburse every landowner who is denied rezoning for a high rise apartment or a shopping centre, and our traditional basis of zoning would be shattered" (1972, 6–7).

With Pearson's information in hand, Williams concluded that any attempt to compensate lost development value would cripple the province financially. Williams, also with experience as a municipal planner where zoning was accepted as a right of the government, asked Pearson to draft a memorandum rejecting the purchase of farmland or development rights because of their fundamental violation of development rights vested in the Crown.

To appease Stupich and create a far less costly proposal, Williams suggested consideration of guarantees of a protected market, cost of production, price support, et cetera; social programs such as a special farmer pension; and rebate taxes on unearned increments phased over several years (Petter 1985, 14–21).

As the Williams version of the bill went through several drafts, William Lane, a municipal lawyer with considerable experience in land-use law (and future first chairman of the Land Commission), was asked to contribute to the drafting process. Lane made several important contributions to the revised bill. Of note was inclusion of the Land Commission in the process of designating the ALR and freeing it from the statutory definition of "agricultural land." On the issue of purchase versus zoning, Petter (1985) relates that to Lane, who like Pearson and Williams had experience with municipal planning, the idea of a province-wide zoning scheme did not seem at all unorthodox, and the notion that such a scheme should require compensation did not even enter his mind. The establishment of provincial zoning in favour of agriculture was simply the "provincial retrieval of provincial powers" (Rawson 1976, 42).

THE LAND COMMISSION ACT

The Land Commission Act was enacted on 18 April 1973, and the commission (of at least five members) was appointed by cabinet in May. The primary role of the commission was to preserve agricultural land. The original act, however, also gave the commission the objectives of establishing greenbelt, land bank, and parkland reserves along with Agricultural Land Reserves.

A source of early misunderstanding was the fact that, in the case of greenbelt, land bank, and parkland reserves, the commission had to purchase (or otherwise acquire) these lands before designating them. Only the ALR could be designated without purchase using traditional zoning tools. Although the commission's responsibilities concerning greenbelt, land bank, and parkland reserves were eventually removed (1977), they served to foreshadow the importance of functional integration and a regional perspective, which would form hallmarks of the emerging regional growth strategy in the Greater Vancouver and other regional districts.

The process for designating the ALR was set out in the act in some detail. Application processes were established to exclude land from the ALR and allow for the subdivision or non-farm use of land in the ALR. The act also included provision for appeal to the Environment and Land Use Committee of cabinet for persons dissatisfied with a decision of the commission.

Permitted land uses in the ALR were outlined and largely restricted to farming. Landowners aggrieved by a decision of the commission could appeal only on a question of law or excess jurisdiction to the Supreme Court. The act specifically stated that land shall be deemed not to be taken or injuriously affected by reason of designation in the ALR. Originally, the act was subject only to two other pieces of legislation, the Environment and Land Use Act and the Pollution Control Act (later the Interpretations Act was added), which speaks to the strength of the legislation.

The relationship of the act to local government plans and bylaws was outlined. With few exceptions, applications under the act were routed first to local governments for advice, and in some cases applications had to be authorized by the regional board or council to proceed to the commission. The act did not impair the validity of municipal or regional district bylaws related to the use of land in the ALR except when there was an inconsistency, in which case the act, regulations, or orders of the commission prevailed, and the inconsistent part of the bylaw was suspended and of no effect. A local government bylaw, such as a building lot line set back, however, could provide regulations in addition to restrictions imposed by the act.

DESIGNATING THE ALR

The first and daunting task facing the Agricultural Land Commission was the establishment of the ALR.[4] The designation process required engagement of the commission with local governments and other provincial agencies.

One of the early criticisms of Bill 42, the precursor to the act, was a lack of local government involvement in establishing the ALR. This omission was corrected by the inclusion, in Section 8 of the act, of a clear and central role for local governments.

This was accomplished by tasking each of the province's twenty-eight regional districts with producing, with assistance from the commission, an ALR plan for the region and subsequently adopting the plan by bylaw. The regional districts in turn worked with their member municipalities in the development of each ALR plan. The act gave regional districts ninety days to complete their work, a time period that the commission could and did extend as required. The decision by the province to directly involve local governments from the outset was inspired and farsighted. It demonstrated faith in what, in 1973, were relatively new entities on the land-use planning/government scene and acted to energize and give purpose to regional districts. In hindsight, to have isolated local governments from development of the ALR would have undermined the farmland preservation program in the long run.

Although local governments insisted on being involved in the designation process, they might not have been fully aware of what they had bought into. As Rawson (1976) reflected with admiration, drawing up the ALR proposals was intense work, particularly for staff at the regional district, municipal, and commission levels. Rawson added that local government staff members performed uniformly well. Through engagement of local governments in the designation process, there was initially a greater sense of local ownership of the ALR plans. The staff who worked so hard to develop the original ALR plans were normally the same personnel engaged with the commission in administration of the act for years to come. These early gains were gradually lost, however, as new local elected government officials, without an appreciation of the historical context of the designation process, came to regard the ALR as strictly a "provincial" reserve.

Clearly, the most important, indeed critical, tool at the disposal of the commission aiding the designation of the ALR was the CLI classification system of agricultural capability. Runka (1977, 136) summarizes the significance of this tool:

> One important decision was that . . . [the ALR] should be based on biophysical parameters, the natural characteristics of the landscape, rather than the variables of market and socioeconomic considerations. Initially, therefore, we had to decide on a technical base that would weather all storms, politically and otherwise, and be as fair as possible to everyone. The Canada Land Inventory (CLI) agricultural capability interpretation, derived from basic soil and climate data, was the only uniform province-wide classification of the land resource available at the time—a very necessary requirement in order to fairly and equitably apply province-wide zoning. Without this basic biophysical inventory, the scheme of credible agricultural zoning intended to preserve agricultural land in the long term would have been very difficult, if not impossible, to implement.

Basing the ALR upon biophysical rather than short-term market and economic conditions has been a perspective that future commissions have also drawn on in administration of the reserve.

To aid regional districts in their work, the process of ALR designation began with input from the Department of Agriculture, which prepared suggested ALR maps. This was also a critical input. The maps drew on the department's overall knowledge of farming in the province and identified lands with soil and climate combinations to support agriculture that were not already urbanized or irreversibly alienated. These maps were generalized second-stage interpretations of basic soil survey and CLI data combined with proposed urban expansion areas on lower-capability or non-agricultural land. The maps would also have given local governments much-needed early direction on the anticipated scope of the Agricultural Land Reserve.

The draft maps were then forwarded to each regional district for consideration. The regional districts consulted with their member municipalities, and public hearings were required to consider and gain input on the proposed ALR. Runka (1977) notes that about 300 information meetings and public hearings were held in the regional districts to allow public participation in drawing the ALR. The regional districts were then required to adopt their ALR plans by bylaw and file them with the commission.

When advising on the ALR, the commission requested that local governments take into account "in-stream" development to ensure that sufficient land was available to accommodate five years of future urban growth. This was to allow for a period of transition and give local governments time to organize their community plans and servicing programs. The commission hoped that this period of grace would be sufficient for the new priority placed on the protection of agricultural land to become a part of the community's thinking (Rawson 1976). Looking back over forty years, the hoped-for fundamental adjustment by local governments away from the post–Second World War attitude that farmland is "urban land in waiting" has been mixed at best.

Once the proposals were completed, the regional districts filed them with the commission. The commission then reviewed in detail each ALR plan to ensure compliance with the intent of the act. As Runka (1977, 137) notes, "the quality of the plans submitted to the commission varied, depending on the attitudes and directions local governments chose to project. During the . . . review stage, therefore, we attempted to ensure basic technical consistency with the agricultural land reserves throughout the Province."

The act provided an opportunity for the commission to recommend amendments to the ALR plans for eventual cabinet consideration, and this was done in a number of cases. In turn, other government ministries and agencies commented on the ALR plans, with the ELUC having final review prior to cabinet's consideration and approval. Once the ALR plans were approved, the commission could designate them according to regional districts. At this point, the original "farmland freeze" orders under the Environment and Land Use Act were lifted, and zoning under the Land Commission Act was then applied (Runka 1977, 138).

The first ALR plan was designated on 13 February 1974 for the Regional District of Okanagan-Similkameen, eleven months after passage of the act and ten months after appointment of the commission. By the end of 1974, twenty-three of the twenty-eight regional districts had their ALR plans designated. Four more were completed in 1975, and the last ALR plan was designated in December 1976. With the designations complete, only 5 percent of the province's land area warranted placement in the ALR—about 4.7 million hectares.

DRAWING THE BOUNDARIES
Despite the speed with which the ALR was designated, considerable care was taken, and fairness was uppermost on the minds of members of the commission when making recommendations to cabinet on the eventual shape of the reserve.

Several factors influenced the final designations. First, lands included in the original "farmland freeze" gave initial shape to the ALR plans. The act did not limit designation of the ALR to CLI Classes 1-4 lands; in fact, it did not mention the CLI. Thus, the commission was not bound strictly by the criteria of Classes 1-4, a point missed by some critics. Since initiation of the "farmland freeze," there had been an ongoing appeal process for those who felt aggrieved. The results of this appeal process would also have been taken into consideration. The practical experience of the Department of Agriculture would also have been valuable input. In designating the ALR, an effort was made to identify cohesive "agricultural landscapes" and ensure their continuity. The regional district process of ALR plan development was also a fundamental component affecting the final shape of the ALR, including the "five-year urban growth" provision. In determining the ALR boundaries, all lands were considered for potential inclusion. As a result, private lands, provincial and federal Crown lands, and Indian reserves were all considered if their biophysical characteristics warranted inclusion in the ALR. With respect to federal lands, including Indian reserves, the regulations associated with the act did not apply. The ALR plans in these latter cases did indicate resource value.

Despite these influences, as Runka (1977) points out, the ALR was solidly grounded on the CLI agricultural capability rating. This greatly assisted in the commission's desire to achieve fairness and objectivity. It also removed the ALR from the spectre of being nothing more than arbitrary zoning. The commission approached its job with creativity given the diversity of the land base and technical considerations. First, all Classes 1 to 4 lands not irreversibly developed, regardless of ownership or tenure, were included in the ALR. Classes 5 and 6 lands were also included where historical land-use patterns indicated that such lands could be used effectively for agriculture in conjunction with Classes 1 to 4 lands. This often occurred in areas of the province where Classes 5 and 6 rangeland underpinned ranching operations. In addition, in many cases, land with a relatively low capability rating with a low *range* of cropping options can actually be highly suitable for a small range of crops (e.g., cranberries in a boggy area). The commission also included some relatively small areas of Class 7 land where such land might have allowed undesirable intrusion of incompatible uses into the ALR.

The commission faced the problem of transferring the underlying irregular pattern of "natural zoning" provided by the CLI mapping into technical descriptions that would be legally defensible and considered practical by landowners (Rawson 1976). Runka (1977, 138) explains that the process of taking the technical CLI data and converting them into straight-line legal boundaries for the purpose of land registry identification and longer-term administration of the ALR was "a long, tough, frustrating job and the results were not altogether successful. Partly because of this, the credibility of the agricultural land reserves has been questioned, especially by nontechnical people." Indeed, successive commissions have expended considerable effort dealing with such anomalies and inconsistencies, real or perceived. Runka suggests that future users of biophysical information, including those generating the data, should be more aware of the problems between natural and legal boundaries.

Another practical problem was the lack of detailed maps or aerial photos in some parts of the province. This assured the commission years of future effort in ALR review processes where these problems were apparent. In addition, particularly given Runka's first-hand experience as an agrologist who worked on development and application of the CLI agricultural capability ratings in British Columbia, there was awareness early on that within some areas (portions of Vancouver Island serve as an example) the CLI information could be improved. But as Runka (1977, 138) reflects, there was an urgency to establish the ALR, and all of the associated problems were recognized, and subsequent commissions have used various techniques to refine the ALR boundaries where appropriate.

In 1978, the Select Standing Committee on Agriculture assessed the CLI agricultural capability of lands in the ALR. Table 4.2 summarizes their work and indicates that the commission could secure 70 percent of the province's prime (Classes 1–3) land within the ALR. The dearth of higher-quality agricultural land in the province is further emphasized when one considers that more water (ditches, channels, streams) is mapped into the ALR than land having a Class 1 rating, and over 40 percent of the ALR has a Class 5 or 6 rating, suitable only for perennial forage crops or natural grazing.

THE COMMISSION

For twenty-seven years, the program was administered by a five- to seven-member "provincial" commission that normally met five days each month. Seventy-two individuals are serving, or have served, on the commission, with an average term of three years. Although fluctuating over the years, the program has usually had a staff complement of twenty to thirty personnel.

Table 4.2. Total CLI agriculturally classified and ALR lands in British Columbia (hectares).

CLI AGRICULTURAL CLASSIFICATION	TOTAL AREA CLASSIFIED	LAND IN THE ALR	THE ALR AS A PERCENTAGE OF LANDS CLASSIFIED
CLASS 1	69,989	52,920	75.6
CLASS 2	397,634	289,079	72.7
CLASS 3	999,644	692,090	69.2
CLASS 4	2,131,581	1,409,080	66.1
CLASS 5	6,137,470	1,468,100	23.9
CLASS 6	5,357,781	431,560	8.1
CLASS 7	14,898,572	167,540	1.1
WATER	-	88,890	-
TOTAL	29,992,671	4,599,259	-

Source: Select Standing Committee on Agriculture (1978).

In April 2000, the commission underwent a fundamental reorganization. The first change was that the province's Forest Land Commission and Agricultural Land Commission were amalgamated into the Land Reserve Commission. To provide a greater regional focus, the traditional five- to seven-member commission was expanded to include a chair and nine commissioners. The commissioners were divided into three regional panels, each with full decision-making power. Each panel was responsible for two of six major regional divisions of the province. The panels maintained consistency by normally meeting for two days a month together and three days in their respective regions.

The second change occurred after the Liberals formed the government in May 2001. The commission remained in place until November 2001, when it was replaced by an interim, five-person, provincial government commission. On 1 May 2002, the commission was again reorganized. A nineteen-person commission (chair and eighteen commissioners) was appointed and divided into six regional, three-person panels, with each member drawn from the region for which the panel was responsible. With responsibility for one regional area, the panels normally meet in the region for three days every other month. Once a year all commissioners meet to discuss broad policy issues. In November 2002, the act was changed (back to the Agricultural Land Commission Act), with removal of the Forest Land Reserve duties, and the commission again focused solely on the preservation of agricultural land.

The most recent structural change to the commission was meant to accomplish the government's "new era" commitment of a more regionally responsive Agricultural Land Commission. This also involved greater deregulation, an "across government" commitment, and encouragement from the provincial government that the commission delegate decision-making power to local governments for subdivision and non-farm-use applications. To date there has been reluctance among many local governments to take on ALR decision making, and only two delegation agreements are in place (the Regional Districts of Fraser–Fort George and East Kootenay).

Despite the stated goal of ensuring that the commission is more regionally responsive, there is little if any evidence that it has not balanced provincial interests with regional wishes. Indeed, the commission can point to a lengthy list of decisions made at the request of local governments and provincial ministries that responded to legitimate local or regional needs.

The six-panel structure and the push for devolving decision making were viewed warily in some quarters, such as by West Coast Environmental Law, Smart Growth BC, and several Vancouver Island regional district planners. Although the commission historically recognized regional differences, this recognition was balanced with a desire to achieve decision-making consistency in similar situations. There was concern that the current structure could lessen consistency and harm integrity. The ability to persuade in a manner contrary to principles of the mandate could be easier with a three-person, rather than a five- to seven-person, commission. One concern during the establishment of the program was a local government's tendency to yield to pressure to convert agricultural land to non-farm use. The structural change to "regional panels" can be viewed as an abandonment of a "provincial" land commission. The decision appears to have been motivated by a desire to move back to a stronger local decision-making model and away from the overview provided by a provincial land commission.

However, until recently, the commission has shown considerable resilience. Regardless of which political party has been responsible for commission appointments, of the fifty-three individuals who have completed service on the commission, rarely has there been a commissioner who did not seem to fully grasp the essentials of the task at hand. Commission after commission has consistently "risen to the mandate." Despite changes in emphasis that have come with these structural changes, the commission's mandate, as laid out in the act, remains basically unchanged.

MODIFICATIONS TO THE MANDATE

Over forty-one years, the mandate of the commission has been amended but remains unwavering in its primary objective to preserve the province's farmland for agricultural use. Key modifications are highlighted below.

1977: A Focus on the Preservation of Agricultural Land

- Responsibility for land banks, greenbelts, and parkland reserves was removed from the act.
- The name changed from Land Commission Act to Agricultural Land Commission Act.
- The minister responsible for the commission could grant landowners leave to appeal a decision of the commission on exclusion applications to the Environment and Land Use Committee of cabinet.
- The Soil Conservation Act was passed and the commission given responsibility for its administration.

1988: Golf Courses an Outright Use in the ALR

- Although initially golf courses were an outright use in the ALR, this use was revoked in the early 1980s. Via a change in regulations, however, golf courses once again became an outright use in the ALR (setting off a groundswell of golf course proposals in the ALR).

1992-94: A Time of Renewal

- A moratorium was imposed on golf courses in the ALR, and once again they require approval from the commission.
- The ability to appeal commission decisions on exclusion applications to ELUC was rescinded.
- The decision-making power of cabinet on applications for exclusion by local governments was given to the commission.
- If an application was declared by cabinet to be in the provincial interest, then it was removed from the commission, and the decision was made by cabinet. (This power has been used only once in eighteen years.)

- An objective was added to the commission's mandate to encourage local governments, First Nations, the provincial government, and its agents to enable and accommodate farm use of agricultural land and uses compatible with agriculture in their plans, bylaws, and policies (ALC Act, Sec. 6[c]).
- The commission was given the opportunity to enter into agreements with local governments to delegate decision-making authority for applications involving subdivision and non-farm use of land in the ALR.
- The relationship between the ALR and local plans and bylaws was clarified and strengthened. Local governments had to ensure that their bylaws were consistent with the act, regulations, and orders of the commission, and any inconsistency was of no force or effect.
- The Municipal Act (now the Local Government Act) was amended to require local governments to forward to the commission official community plans prior to adoption for comment.
- The Forest Land Reserve was designated and the Forest Land Commission appointed. Staff of the Agricultural Land Commission provided operational support to the Forest Land Commission.

1999: The Provincial Interest Defined

- After completion of a report by Moura Quayle (1998) regarding the provincial interest and the ALR, the legislation was amended (and came into effect in 2000) to include language defining the provincial interest.

2000: Commission Amalgamation

- The Forest Land Commission and the Agricultural Land Commission were amalgamated into the Land Reserve Commission.
- The commission was reorganized into three panels comprised of three persons each.

The Farmland Preservation Program in British Columbia

2002: The Agricultural Land Commission

- The Soil Conservation Act that the commission administered was repealed, which in turn assisted in reducing the commission's application workload.
- The Forest Land Reserve duties were removed, with the commission again focused solely on its mandate to preserve agricultural land.
- The commission was restructured (a chair and eighteen commissioners) into six regional, three-person panels, with each member drawn from the region for which the panel was responsible.
- The Soil Conservation Act, which the commission had administered, was rescinded, which in turn assisted in reducing the commission's application workload.

2014: The Agricultural Land Commission

- On-going reviews continue to impact how the commission and ALR operate.

Two Key Duties

As delineation of a 4.7 million-hectare agricultural zone was finalized, two roles quickly emerged and have been central to the commission's administration of the ALR for three decades. First, under the act, processes were established to allow applications. This quasi-judicial role of deliberating on applications has been at the core of the commission's day-to-day life. Second, the review of local government land-use policy, plans, and bylaws emerged as another cornerstone activity.

THE APPLICATION BOX

It seems unlikely that those who crafted the original Land Commission Act fully grasped the impact on the commission's future workload of the application review process. The commission has dealt with nearly 40,000 applications under the act, an average of about 1,000 a year. Although approximately 25 percent of new applications generate requests for reconsideration if approval is denied, there has been a notable decline in applications, particularly after the initial ten-year period, as noted in Table 4.3. The decline in applications, however, can be attributed partially to more permissive regulations for certain non-farm uses within the ALR (e.g., home occupation and bed-and-breakfast accommodation).

Table 4.3. ALR application activity, 1976–2010.

PERIOD	NUMBER OF APPLICATIONS (ALL TYPES)	AVERAGE NUMBER OF APPLICATIONS PER YEAR
1976–1985	17,357	1,736
1986–1995	9,746	975
1996–2005	5,101	510
2006–2010	2,947	589

Source: PALC (2012).

The routine job of dealing with applications has evolved into the core activity of the commission. There are basically two types of applications: those that alter the boundaries of the ALR (exclusions and inclusions) and those that propose the subdivision or non-farm use of land within the reserve. About 90 percent of all applications are initiated by private landowners, with the remainder coming from local governments, the commission, or other provincial agencies. The ability of landowners to make applications directly to the commission has contributed to security of the program by acting as a "pressure valve," dissipating, to a degree, concerns over having their lands originally designated in the ALR.

With the considerable experience of the commission, the application process has been finely tuned over the years to ensure thoroughness, fairness, and efficiency. Because of the varying complexity of applications, the review process can take three months or longer. The application files have always been open to scrutiny, and, using electronic means, the commission is moving toward ensuring greater transparency of the process.

THE ROLE OF LOCAL GOVERNMENTS

Most applications have to be submitted first to the municipalities or regional districts within which they originate. In the case of exclusion, applicants must also provide public notification of the application. A minor number of applications (e.g., highway widening) are routed directly to the commission, but even in these cases local governments will be informed of decisions, and the commission might seek the advice of the local government in question.

When a local government (or the commission) initiates an inclusion or exclusion application, a public hearing is required. In the case of landowner applications for inclusion or exclusion, the local government might hold a public information meeting if considered appropriate.

Local governments have developed their own ALR application review procedures, and many have developed materials to assist the public in understanding the process. In most cases, the act requires local governments to authorize forwarding of an application to the commission. Authorization provides local governments with an opportunity to ensure the integrity of their official community plans and bylaws. If the application is not authorized, then the process ends. Historically, the use of authorization has varied among local governments, with several automatically (by policy) forwarding all applications to the commission regardless of whether the proposal is contrary to local land-use regulations or not. Other local governments are far more active in using their power of authorization. To further ensure the integrity of local bylaws, on every decision letter that involves an approval, the applicant is informed that an approval by the commission does not usurp local government bylaws, which must still be observed.

With each application authorized, the local government provides a variety of background documentation and a recommendation to the commission. The commission considers, but is not bound by, the recommendation. The regulations also set out an application fee structure, with approximately half of the amount retained by the local government.

THE APPLICATION REVIEW PROCESS

Once the commission receives an application, it is logged into its "application tracking system." Given that the commission has dealt with about 40,000 applications to date, an electronic tracking system has become an administrative imperative. Research staff have responsibilities to gain familiarity with particular areas of the province. The research officer reviews the material submitted by the applicant, along with the documentation and recommendations supplied by the local government, and prepares a report for the commission. The commission attempts to examine as many applications on site as possible.

The commission's review of an application is comprehensive and strives for fairness, and applications are routinely reconsidered when justified by new information. The review considers site-specific details and broader issues. The "ingredients" of a review include the proposal and reasons for it, CLI data, previous applications on the subject property and properties in the vicinity, relevant policies of the commission, history of agricultural use of the site, investments in agricultural infrastructure in the area, the property's context within the larger agricultural community, potential off-site impacts, the possibility of buffering to mitigate impacts if approved, potential for the proposal to create expectations if approved, input provided by the local government, and relevant zoning, official community plan, and regional growth strategy information.

In the case of exclusion applications, the commission must give prior written notice (to the applicant, local government, and possibly other parties) of the meeting to consider the application. The applicant is provided with a copy of application material prior to the meeting. At the meeting, the commission can accept written submissions and hear the applicant and other interested parties. It has become a standard procedure of the commission to give the applicant an opportunity to speak directly to the commission on the proposed exclusion.

FOUR DECADES OF APPLICATIONS[5]

Despite the continuance of applications, there is no conclusive evidence that the process has undermined objectives of the farmland preservation program. But the applications can overwhelm the commission and sap its energy. They have diverted the commission away from spending more time on analytical tasks, awareness building, and proactive planning.

Runka (1977) predicted when the ALR was first designated that there would be an ongoing need to fine-tune its boundaries. After forty years, 141,000 hectares have been excluded from the reserve, and nearly 184,000 hectares have been included. Taking all exclusions into account, the original ALR has been reduced by only 3 percent. In overall land area, inclusions have more than made up for exclusions. This has resulted in a larger, not a smaller, ALR after four decades.

Quality, however, has suffered to a degree. Most of the land included in the ALR (89 percent) has been in the northern two-thirds of the province, with only a small percentage being prime agricultural land (CLI Classes 1–3). The result is that, for every hectare of prime land included in the ALR, three hectares of prime land have been excluded. Prior to introduction of the farmland preservation program, it was estimated that from 4,000 to 6,000 hectares of prime farmland were lost annually to urbanization (PALC 1983). Over the past four decades, this loss of prime agricultural land has been slowed to about 600 hectares excluded from the ALR each year.

The most prevalent application is for subdivision and non-farm use. Commission statistics show a high propensity to approve applications of this type (e.g., 71 percent between 1981 and 2000). There is an inherent danger in rushing to a judgment on the basis of such evidence. One would have to review applications for subdivision and non-farm use in some detail to gain a true picture of the quality of decision making. Often an application involves the subdivision of a property along the ALR boundary or a minor non-farm use that has little bearing on the productive value of the property. Unfortunately, there are few resources to do close examinations of this type of application activity.

Where evidence does exist, it points to the commission's slowing considerably the subdivision of ALR land. One internal review examined the first eighteen years of commission decision making in Delta, a municipality south of Vancouver. It indicated that on average only one additional legal parcel was created in the ALR each year over this period. The study also demonstrated a consistent decision-making pattern of never allowing the splitting of large farm parcels into smaller units. All of the applications approved involved small lots or resulted from the commission's home site severance policy, which provides an opportunity for farmers to sever a small parcel for a home site upon retirement.

A study of ALR subdivision activity in the Vancouver Island community of North Cowichan considered agricultural criteria as a basis for requesting subdivision. Only 9 percent of applicants sought subdivision in order to improve farm operations. In contrast, 80 percent of applicants were motivated by personal reasons, usually to sever a building site for a family member.

THE NEXT STEPS

The commission expends considerable energy on what is often negative rather than positive administration for those whose primary intention is to take action contrary to farmland preservation. Most applications play out the classic dilemma of balancing individual desires and community values. There has been a gradual reduction in the number of annual applications to the commission over the past forty years; however, any gains that might have been realized in freeing staff resources and commission time for other efforts in support of the program have been lost to staff reductions. Provincial governments have often viewed the application process as the sole legitimate role of the commission, placing it in an application box from which it has been difficult to escape. Fortunately, the commission has worked to broaden its role in several ways, including considerable effort to positively influence land-use-planning processes.

To further this end, in November 2010, the chair of the Agricultural Land Commission completed an extensive (113-page) *Review of the Agricultural Land Commission*. The review contained eight key recommendations, one of which was "that the work of the ALC be repositioned away from being reactive and focussed on applications, to a proactive planning model that will enable it to strengthen ties to local government land use planning, deal with emerging issues as they relate to agriculture, and undertake ALR boundary reviews" (Bullock 2010, 7). This recommendation was clarified by the ALC chair in August 2012 in outlining a desire to adjust commission priorities to lessen resources expended on reactive activities associated with applications, which have been at the expense of other important parts of the ALC mandate. The goal is to reduce the commission's budget devoted to application work from the current 80 percent to 30 percent. This shift will give the commission a greater opportunity to work proactively on ALR boundary reviews, enforcement, and engagement in planning and policy processes.

PLANNING FOR AGRICULTURE

From the early days of the program, the commission has had a small policy and planning branch, staffed by planning professionals, to assist the commission in its work with local governments and provincial ministries and agencies. This facet of its work has centred on the review of an array of land-use policies, planning documents, and bylaws. Although the commission maintains a close working relationship with resource ministries and other agencies, the largest part of its effort is working with local governments in the review of plans and bylaws. The legislative backstop to this work is comprised of the following:

1. the commission's mandate to encourage local governments and others to accommodate farming in the ALR in their plans, bylaws, and policies (ALC Act, Section 6[c]);

2. the need to ensure consistency between local government bylaws and the Agricultural Land Commission Act, regulations, and orders of the commission (ALC Act, Section 46); and

3. the requirement that local governments forward copies of draft official community plans to the commission for comment prior to approval (Local Government Act, Section 882[3][c]).

For the commission, working with local governments following imposition of the ALR has been challenging, calling on a balance of steadfastness and diplomacy. Passage of the Land Commission Act did not, and was never intended to, remove a local government's plan and bylaw authority. Hence, both the commission and local governments were charged with responsibilities for zoning and regulating the same land base—the ALR. Considerable effort has been expended to sort out the land management relationship.

Beyond legislative requirements, the commission's interest in land management processes is grounded on several realities. Plans and bylaws inconsistent with farmland preservation objectives have had considerable undermining potential, setting off expectations of land-use change in farm areas. At the same time, when local governments have made policy adjustments that effectively plan for and secure agriculture's place in their communities, doing so can shift the focus of the urban development industry. Plans have also been used as the vehicle of choice in the pursuit of adjustments to the ALR. Finally, the commission takes an active interest in evolving urban land-use policy, recognizing fully that in many areas the ALR boundary is the obverse of an urban growth boundary. The commission recognizes that failures in urban planning will bear directly on its efforts in pursuing provincial farmland preservation objectives.

A PERIOD OF REACTION

During the 1970s and into the mid-1990s, the commission found many official community plans displaying indifference to agricultural issues. It was clear that urban concerns dominated planning processes. Two local government processes emerged in the first half of the 1990s, however, that reset the bar: the Langley Rural Plan (1993) and the Greater Vancouver Regional District (GVRD) Livable Region Strategic Plan (1996).

The Township of Langley, located in the eastern portion of the GVRD (now known as Metro Vancouver), contains 23,500 hectares of land in the ALR, 77 percent of the land base of the municipality. The Langley Rural Plan involved a process designed not to be dominated by urban issues. The plan was a watershed municipal document that created a sub-area plan focused on a rural, predominantly farming, area. The planning process was inclusive; drawing on the farm community for input, it included a series of economic policies geared to agriculture and placed farming as the priority use in the plan area.

Metro Vancouver, stretching back to the Lower Mainland Regional Planning Board of the 1950s and 1960s, has demonstrated a consistent and keen awareness of the implications of urban growth for farmland preservation. Agriculture is a major land-use activity in the metropolitan area, generating gross farm receipts in 2010 of $789.5 million, 27 percent of the province's total (Statistics Canada 2011). Metro Vancouver's Livable Region Strategic Plan broke with traditional approaches to projecting population trends into land-use needs. It turned the process on its head by identifying and protecting what was most important to the people of the region first. The result was the definition of a green zone that includes major parks, watersheds, environmentally significant features, and working forests and agricultural lands. The ALR represents 26 percent of the region's green zone. Each Metro Vancouver municipality has designated its portion of the green zone lands, and any substantial change triggers an amendment to the plan that requires regional board approval. Since 1996, the ALR in the most heavily populated region of the province has decreased by only 1.3 percent, a period during which the region's population grew by over 481,000 or 26 percent.

For the commission, Langley and Metro Vancouver provided models at the municipal and regional levels that regarded agriculture as a full partner in their planning processes.

MOVING BEYOND PRESERVATION

Frustrated by official community plans that did not effectively grapple with issues important in farm areas and that failed to adequately engage agriculture as a full partner in growth management debates, the commission decided to build upon the Langley Rural Plan as a model that could have broader application. The commission also recognized that planning and agriculture have not traditionally been strong examples of crossover disciplines. In response, the commission published *Planning for Agriculture: Resource Materials* (Smith 1998) to help bridge the gap between land-use planning and agriculture. The document was aimed most directly at municipal and regional districts engaged in plan and bylaw development and represented a call to move agriculture into the planning mainstream at the local level.

The basic principles of *Planning for Agriculture* are livability, agricultural security, land-use compatibility, and stability to ensure continuing industry confidence. The challenge was to reverse decades of auto-oriented suburban sprawl onto farmland, where the conversion of agricultural land to non-farm uses was considered not only inevitable but also natural.

With populations growing, and in anticipation of continuing demand for urban growth on farmland, the commission thought that establishing planning policies designed to secure agriculture's place in communities could act as a counterbalance. It could play a part in forcing a rethinking of urban form and the pursuit of Smart Growth principles. Some of the key recommendations of *Planning for Agriculture* focused on farm-inclusive processes, agricultural area plans, and linear planning processes ("edge planning") along urban and agricultural interfaces. The ALR offered a clear point on which to "hang" edge planning to ensure greater land-use compatibility and permanence. The major thrust of agricultural area planning is to provide a vehicle that can both identify and offer means to deal effectively with agricultural issues in greater detail than can traditional official community plans. These focused planning exercises allow an opportunity to examine land-use dynamics inside the ALR, engage farmers, and develop results-based policies. Given society's gradual disconnection from agriculture (in 2006 only 1.5 percent of all British Columbians lived on farms), planning for agriculture also offers the potential for awareness building and reconnecting.

British Columbia enacted right-to-farm legislation in 1996. At the same time, amendments were made to the Local Government Act and Land Title Act aimed at giving local governments new tools to plan for agriculture. The "right to farm" and "planning for agriculture" package was rolled into the Strengthening Farming" program. This program has been implemented jointly by the Ministry of Agriculture and the Agricultural Land Commission, with program coordination handled by the ministry. The program is designed to augment, not replace, the long-standing commission work with local governments on plan and bylaw reviews.

Although the commission has always had planning staff with regional responsibilities, the ministry underwent a significant shift of staff resources. Its core Strengthening Farming coordinating staff were augmented with about fourteen regional agrologists who devote a portion of their time to the Strengthening Farming program. Ministry agrologists and commission planners have formed provincial "agri-teams." Each team is assigned to specific local governments to work "on call" on a variety of local agricultural issues. This alliance draws together planning and agricultural talents and helps to achieve a crossover of the two disciplines to the benefit of both.

The suite of legislative actions to support the Strengthening Farming initiative included a stronger emphasis in the Local Government Act (LGA) on the need for official community plan policies that maintain and enhance farming in the ALR. Provision has been made to designate "development permit areas" in official community plans for the protection of farming by encouraging buffering and more "farm friendly" urban design next to agriculture. Officers can refuse approval if a subdivision next to a farming area does not include sufficient buffering or if unnecessary road endings are directed toward the ALR. The LGA provided for "farm bylaws" that could specifically deal with issues such as the siting of farm buildings and operational techniques.

Farm bylaws require approval of the minister of agriculture before adoption. In addition, the minister has the power to develop bylaw standards to assist in updating agricultural portions of zoning bylaws housed in a *Guide for Bylaw Development in Farming Areas* (MAF 1998). The province can also require a review of zoning bylaws affecting the ALR if there is a concern about their impacts on agriculture. Like agricultural area plans, these legislative tools were designed to provide additional opportunities to focus on agriculture and enhance security of the farmland base—and thus have a direct and mutually complementary connection to the farmland preservation program.

Local governments across British Columbia have been reconnecting with their farm communities more directly and taking up the challenges of planning for agriculture. Forty-eight local governments have the benefit of an "agricultural advisory committee" to provide advice to councils and regional boards on a wide range of issues. Following in Langley's footsteps, forty-two agricultural area plans and strategies have been completed, and a further thirty-one are under way or being considered by communities throughout the province.

To further emphasize the relationship between local government plans and bylaws, the Agricultural Land Commission published the *ALR and Community Planning Guidelines* (2004). This guide, along with several others, including a *Guide to Edge Planning* (Ministry of Agriculture 2015), and the application of land-use inventories and Geographic Information Systems (GIS) in farm areas, have been developed for local governments to assist in drafting plans and bylaws. Again they point to the importance that the commission places on local land-use policies working in partnership with the provincial farmland preservation program. Taken together, the commission, and now Ministry of Agriculture, efforts are designed to build partnerships among the farm community, local governments, and the province and to move beyond the single issue of farmland preservation.

SUMMING UP

By any measure, British Columbia's program to safeguard scarce farmland resources has met its preservation objective and significantly reduced the conversion of farmland in the province to urban and other non-farm uses. The estimated loss of as much as 6,000 hectares of prime agricultural land annually has been reduced, on average, to about 600 hectares since the establishment of the ALR.

In many BC communities, the ALR has influenced urban growth policy directed at hillside and more compact development alternatives. Within Metro Vancouver, which accounts for over 52 percent of the BC population, the ALR has worked in harmony with the region's growth strategies. The farmland preservation program has provided a building block that has supported the region's livability strategy aimed at developing compact communities. The strategy also protects a designated green zone, which includes agricultural and parkland, watersheds, and environmentally important areas. The result has been the protection of foodlands *within* the province's most heavily urbanized area, contributing to the economy and a local food source.

As in the past, the program will undoubtedly change to meet new challenges. The ALR and work of the commission have enjoyed consistently strong public support. Attacks on the ALR tend only to solidify support. An opinion survey in 1997 found that over 80 percent of British Columbians considered it unacceptable to remove land from the ALR for urban uses, and a survey done in 2008 found that 95 percent of British Columbians supported the ALR and its farmland preservation policies.

Parallel to the farmland preservation program is the need to ensure an economically healthy and environmentally sustainable agricultural sector. But a strong agricultural industry alone will not be enough to thwart the urban/industrial conversion of farmland. A key litmus test of the program will be how successful the commission is at ending the perception that its role is that of a rationing board, slowly but surely meting out the province's farmland base to alternative uses. Successfully instilling a land management ethic that recognizes farmland preservation as a social value and the ALR as a treasured and permanent part of the landscape will take the constant and active support of successive provincial governments. It will also take local growth management policies founded on the view that the best and highest use of agricultural land is agriculture—now and in the future. The ALR rests on land-use scales to provide balance against competing uses: "Without the courage to hold firm, with stakes in the ground, there will be no incentive to better manage our land base in the face of competing uses. We must halt the slow but steady erosion of our agriculture and food resources, and support our varied agricultural industries. As a forward thinking society, we must dig in, take responsibility, and make sure that future generations have a vibrant agricultural land base" (Quayle 1998, 27). *Holding firm* has been the Agricultural Land Commission's challenge, a challenge now being shared.

NOTES

1   At the time of writing, the BC government was reviewing the role of the Agricultural Land Reserve program and proposing major changes.

2   At the time of the 1911 census, British Columbia and Ontario were the only provinces with more than 50 percent of their populations living in urban settings (Department of the Interior Canada 1915, 95). At the time of the 2011 census, British Columbia and Ontario were the most urbanized provinces at 86.2 percent and 86.0 percent, respectively, compared with 81.1 percent for the country as a whole. In 1971, 60.2 percent of the population of British Columbia was concentrated in the southwestern portion, a figure that rose to 72.2 percent by 2011 (BC Stats; Statistics Canada 2011).

3   The 1963 university textbook *Regional and Resource Planning in Canada* exposed students to Ralph Krueger's passionate concern over the disappearance of the Niagara Fruit Belt. Within this book A.D. Crerar reviewed the loss of farmland in metropolitan regions of Canada, and Norman Pearson wrote about trends of urbanization. Pearson later wrote about the impacts of urban and recreational planning (1972). Both Crerar and Pearson would play influential roles in the development of the Land Commission Act.

4   This summary of the original ALR designation process draws heavily from two sources, both with direct involvement in it. One is a paper written by the first general manager and former chair of the commission, Gary Runka, "British Columbia's Agricultural Land Preservation Program" (1977). The second is a report written by Mary Rawson, one of the original commissioners, *Ill Fares the Land* (1976). Both sources are recommended for persons interested in a first-hand look at the beginnings of the ALR.

5   I have compiled all statistical data from the files, reports, and website of PALC.

REFERENCES

Agricultural Land Commission. 2004. *ALR & Community Planning Guidelines*. British Columbia: Provincial Agricultural Land Commission.

Baxter, D. 1974. "The British Columbia Land Commission Act: A Review." Master's thesis, University of British Columbia.

BC Stats. http://bcstats.gov.bc.ca/Home.aspx.

British Columbia New Democratic Party. 1972. *N.D.P.: A New Deal for People*. Vancouver: Allied Printing Trades Council.

Brundtland, G.H. 1987. *Our Common Future*. Report of the World Commission on Environment and Development. Oxford: Oxford University Press.

Bullock, R. 2010. *Review of the Agricultural Land Commission*. Province of British Columbia: Provincial Agricultural Land Commission.

Cannings, R., and S. Cannings. 1996. *British Columbia: A Natural History*. Vancouver: Greystone Books.

Department of the Interior Canada. 1915. *Atlas of Canada—1915*. Ottawa: Government of Canada.

Hodge, G., and L.M. Robinson. 2001. *Planning Canadian Regions*. Vancouver: UBC Press.

Krueger, R.R. 1963. *Regional and Resource Planning in Canada*. Toronto: Holt, Rinehart and Winston.

MAF (Ministry of Agriculture and Food). 1998. *Guide for Bylaw Development in Farming Areas*. Abbotsford: Province of British Columbia.

Ministry of Agriculture. 2015. *Guide to Edge Planning*. Abbotsford: Strengthening Farm Program.

PALC (Provincial Agricultural Land Commission). 1973. Internal briefing material on Bill 42, Land Commission Act. PALC files, 8 March.

——. 1983. *Ten Years of Agricultural Land Preservation in British Columbia*. Burnaby: PALC.

——. N.d. "History of the ALR." http://www.landcommission.gov.bc.ca/ publications/Alr_history.htm.

Pearson, N. 1972. "Fraser Valley: Rape It or Preserve It?" Paper presented at the seminar Land Use in the Fraser Valley: Whose Concern?, UBC Centre for Continuing Education and Faculty of Agricultural Sciences, Vancouver.

Peterson, S. 1971. "A Program for the Preservation of Farm Lands in the Fraser Valley Based on Soil Capability Classification." Report, Department of Agriculture, Victoria.

Petter, A. 1985. "Sausage Making in British Columbia's NDP Government: The Creation of the Land Commission Act, August 1972-April 1973." *BC Studies* 65: 3–33.

Province of British Columbia. 1972. Order-in-Council 4483, Environment and Land Use Act, 21 December.

——. 1973. Order-in-Council 157, Environment and Land Use Act, 18 January.

Quayle, M. 1998. *Stakes in the Ground: Provincial Interest in the Agricultural Land Commission Act*. Victoria: Province of British Columbia.

Rawson, M. 1976. *Ill Fares the Land*. Report for the Ministry of State, Urban Affairs. Toronto: Macmillan.

Runka, G. 1977. "British Columbia's Agricultural Land Preservation Program." In *Land Use: Tough Choices in Today's World*. The Proceedings of a National Symposium, 21–24 March, 1977, in Omaha, Nebraska, 135–43. Ankeny, IA: Soil Conservation Society of America.

Select Standing Committee on Agriculture. 1978. *Inventory of Agricultural Land Reserves in British Columbia*. Phase 1 Research Report. Richmond, BC: Select Standing Committee on Agriculture.

Smith, B.E. 1998. *Planning for Agriculture: Resource Materials*. Burnaby: PALC.

Statistics Canada. 2011. *2011 Census of Agriculture*. http://www.statcan.gc.ca/eng/ca2011/index.

Vancouver Sun. 1973. "Farmland Freeze Sparks Vote." *Vancouver Sun*, 3 February.

Wilson, J.W., and J.T. Pierce. 1982. "The Agricultural Land Commission of British Columbia." *Environments* 14 (3): 11–20.

CHAPTER FIVE

Learning about the Agricultural and Food System in Your Municipality

J.C. (JIM) HILEY

THE AGRICULTURAL AND FOOD SYSTEM IN LOCAL AND REGIONAL LAND-USE PLANNING

In simple terms, the agricultural and food system is comprised of six broad sectors (Hiley et al. 2011). Each sector produces specific outputs and is integrated with other aspects of the system through one or several links (see Figure 5.1). For instance, farmers might grow tomatoes and sell them to the secondary sector (e.g., manufacturers or wholesalers), the tertiary sector (e.g., farmers' markets or restaurants), or directly to the consumers sector (e.g., produce stands at the farm gate). Every sector relies on the support services sector for inputs and the waste management sector for controlling outputs yet to realize economic value.

Until the mid-1990s, the agricultural and food system was held as a high-priority land use by the public and reflected as such in the legislated land-use planning process. Citizens have continued to demonstrate their ardent interest in over 2,300 grassroots initiatives in support of local agricultural and food systems (Canadian Co-Operative Association 2008). Their commitment is also expressed in the evolution of a "people-centred" national food policy, a framework to integrate exciting local and regional projects (Food Secure Canada 2011). Separately and collectively, these efforts attempt to enhance the positive economic, social, and environmental contributions of the agricultural and food system to the development of their communities (British Columbia Ministry of Agriculture and Lands 2007, 2008; Reimer 2003). In contrast, the importance of the agricultural and food system in the formal land-use planning process went in the opposite direction in the same period.

Figure 5.1. Sectors comprising the agricultural, food, and agri-products industries.

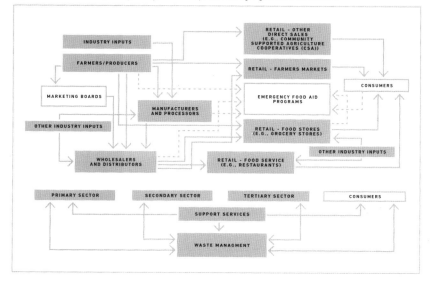

Source: Based upon Harry Cummings and Associates (1999); used with permission.

In general, in the past two decades, provinces selectively delegated their constitutional authority in land-use planning to lower-tier governments without adequate financial and human resources (Federation of Canadian Municipalities 2012). At the same time, provinces provided less direction on policies on priority land-use issues such as the preservation and conservation of prime agricultural land. Notable exceptions to this trend were the formally designated Agricultural Land Reserves in British Columbia and Quebec and, to a much smaller extent, Ontario (Hiley et al. 2011). The ALRs serve a dual purpose in the land-use planning process:

1. They encourage viable farming operations on dependable agricultural land.
2. They act as de facto urban growth boundaries, thereby encouraging compact, fiscally sustainable residential and other forms of urban development (Smith and Haid 2004).

Concomitant with the decline in prominence of the agricultural and food system in the land-use planning process has been the conversion of millions of hectares of dependable farmland to sprawling, highly inefficient forms of non-agricultural development (Hofmann, Filoso, and Schofield 2005). The situation has been made worse by the lack of relevant academic training and the paucity of professional development opportunities for new and established planners with regard to the agricultural and food systems within their jurisdictions (Hiley et al. 2011). Hence, decisions about the future mix of land uses in an area might have been made without sufficient information on possible adverse consequences to the agricultural and food system.

Two other factors will gain prominence in coming decades in efforts to maintain the agricultural and food system as a viable land use. The first factor is demographics: the majority of farm operators are currently in middle age and approaching retirement, with very few young or new farmers filling the growing void. Land is often the most valuable asset to farm families and must be sold or in some other way monetized to finance their departure from the workforce. Not surprisingly, there can be considerable resistance within the farming community to any actual or perceived restriction on farmers' ability to sell land at the highest price, especially if few people are able or willing to take over their operations. The second factor is the size and configuration of any parcel that will remain for agricultural production once the current generation of producers has released their land base. A significant loss of the total area of prime or dependable agricultural land in Canada, as well as considerable fragmentation of the remaining land base, will influence how effectively the agricultural and food system functions in the future.

To summarize, information and knowledge gaps about this system at local and regional levels adversely affect its due consideration in land-use planning. There is an urgent demand to effectively describe the agricultural and food system in a jurisdiction to help

- scope the economic, social, and environmental characteristics of the system at local and regional levels;
- act as a baseline against which to assess its importance relative to other major land uses;
- identify potential contributions to future development; and
- provide a rationale with which to leverage human and financial resources for more in-depth studies.

Therefore, the research objective is to develop and evaluate a method to profile the agricultural and food system in a lower-tier government by integrating existing data and information from diverse sources.

THE APPLIED RESEARCH METHOD

A framework to integrate policies supportive of the agricultural and food system provides the context for the development of a classification scheme. Value chains refer to the mutually beneficial links among all businesses involved in meeting consumer demand for agricultural products (Devanney 2006). The value chain theory is a widely regarded alternative to the industrial agricultural policy in that it recognizes that value is added beyond the production of a commodity by the primary sector within a region (Morton 1995). Initially associated with the farm level, the concept of value chains is now applied to co-located, cooperating firms throughout the agricultural and food system.

With regard to the definition of terms, value is added to a good or service when consumers in the marketplace are thought to be willing to, or actually do, pay a premium for a product altered by an activity (Born and Bachmann 2006). The incremental value added by a good or service need not be reflected in a higher price since product price-setting is a complex process. The scope of activities considered in this applied research includes all sectors of the agricultural and food system.

A two-level, activity-based classification was created following a literature review of twelve national and international classification systems. The First-Order Activity Classes represent ten broad categories, whereas the Second-Order Activity Classes generally have five more groupings of activities (see Table 5.1). Exceptions are the lack of classes for the waste management class, four categories for the wholesaler and distributor trade class, and six categories for the consumers class.

A technical steering committee was formed with membership from a number of organizations within the municipality that use an evidence-based approach to promote viability of the agricultural and food system (Morton Horticultural Associates 2009). To ensure that the researcher is fully aware of the type of data and information required to support these initiatives, representation on the committee includes municipal, regional, provincial, and federal agencies in addition to a major local farm organization, a private agricultural service business, and a member of the public. The committee's key roles are to provide direction and feedback to development of the classification scheme and data analysis as well as interpretation and discussion of results.

Table 5.1. Hierarchical classification of activities comprising the agricultural and food system.

FIRST-ORDER ACTIVITY CLASSES	SECOND-ORDER ACTIVITY CLASSES
PRIMARY AGRICULTURAL PRODUCTION	CONVENTIONAL CROP, LESS CONVENTIONAL CROP, CONVENTIONAL LIVESTOCK, LESS CONVENTIONAL LIVESTOCK, OTHER PRODUCTION
SUPPORT SERVICES TO PRIMARY AGRICULTURAL PRODUCTION	CHEMICAL SALES, CROP SERVICES, IMPLEMENT DEALERS AND SERVICES, LIVESTOCK SERVICES, OTHER SUPPORT SERVICES
MANUFACTURING	CROP-BASED MANUFACTURING, LIVESTOCK-BASED MANUFACTURING, BEVERAGES, AGRICULTURAL CHEMICALS AND IMPLEMENTS, OTHER MANUFACTURING
WHOLESALER (W) AND DISTRIBUTOR (D) TRADE	AGRICULTURAL SUPPLIES, FARM PRODUCTS, FOOD/BEVERAGES/TOBACCO, OTHER W/D TRADE
RETAIL TRADE	NURSERY/FLORIST/GARDEN CENTRE, FOOD/BEVERAGE STORES, FOOD (HEALTH) SUPPLEMENTS, PET FOOD AND RELATED STORES, OTHER RETAIL TRADE
TRANSPORTATION AND WAREHOUSING/STORAGE (W/S)	REFRIGERATED W/S, FARM (CROP) PRODUCTS W/S, FARM (LIVESTOCK) PRODUCTS W/S, AGRICULTURAL AND MANUFACTURING PRODUCTS TRANSPORTATION, OTHER TRANSPORTATION AND W/S
NON-FOOD SERVICES (PROFESSIONAL, SCIENTIFIC, TECHNICAL)	AGRICULTURAL SYSTEM, FOOD SYSTEM, AGRICULTURAL INDUSTRY ASSOCIATION, FOOD SYSTEM INDUSTRY ASSOCIATION, OTHER NON-FOOD SERVICES
ACCOMMODATION AND FOOD SERVICES	FULL-SERVICE RESTAURANTS, LIMITED-SERVICE RESTAURANTS, SPECIAL FOOD SERVICES, DRINKING PLACES, OTHER ACCOMMODATION AND FOOD SERVICES
WASTE MANAGEMENT	NO ACTIVITY CLASSES IDENTIFIED
CONSUMERS	NUTRITIOUS DIET EXPENDITURES FOR FEMALE (19 OR LESS); MALE (19 OR LESS); FEMALE (20–64); MALE (20–64); FEMALE (65 AND OVER); AND MALE (65 AND OVER)

Source: Morton Horticultural Associates (2009). *A Value-Added Strategy for Agriculture: Kings County, Nova Scotia*. FY 2009–2014. Cold Brook, NS: Morton Horticultural Associates.

RESULTS FROM A PILOT STUDY IN KINGS COUNTY, NOVA SCOTIA

Kings County was selected because of a demonstrated commitment to the agricultural and food system spanning hundreds of years. The county, the heart of the Annapolis Valley, has had farming as an established land use for over 250 years. In the nineteenth century, farmers supplied the global market with apples and products made from tree fruits. In the twentieth century, farming was characterized by greater crop and livestock diversity, indicative of changing global supply and demand, the entrepreneurial skills of county farmers, and superior natural capital, such as climate, land, and water resources. Planning for the compatibility of agricultural and other land uses is a major theme early in the twenty-first century.

The importance of the agricultural and food system in the land-use planning process is premised on its many contributions to sustainable development of the county. The key land-use planning document—the Municipal Development Plan—was first adopted in 1979 and refined four times, with the most recent review process initiated in 2011. Throughout this time, the county has worked to increase the viability of farming and other sectors of the agricultural and food system through policies such as agricultural land protection, definition of agricultural districts, and control of the physical form of urban development at the edges of these districts.

An inventory of data and information sources related to the county's agricultural and food system documented sixty-five possible databases, of which sixty databanks could not be accessed for a number of reasons. The reason most often cited was potential concern related to provincial privacy legislation. A second reason was shortage of staff resources when the request was made to provide a summary of a database within the classification scheme. Further, though the study was conducted in consultation with local and provincial government personnel, a federal government agent led the analysis. A possible jurisdictional issue was a third reason raised. These were the main reasons given when access was denied.

Of the five databases to which access was permitted, one database custom-processed from Statistics Canada's *2006 Census of Agriculture* contained detailed descriptions of different types of agricultural operation (Michiels and Hiley 2010). Persons in charge of agricultural operations were to complete a questionnaire if they produced and intended to sell any of the following products: crops, livestock, poultry, animal products, or other agricultural products as defined by the agency (Statistics Canada 2006). Each farm was assigned to one of twelve farming systems on the basis of the dominant source of income from the sale of agricultural products. The study profiled or summarized the inventory of crops and livestock, machinery and equipment, as well as financial and social characteristics for each farming system. The custom database was developed for the entire county.

A second source was the Business Registry Database (BRD) of Statistics Canada. The purpose of the BRD was to provide basic information so that the federal statistical agency could develop reliable economic surveys of the Canadian business community (Statistics Canada 2009a). The business community exerts a significant influence on land use at the local government level through capital investment and financial activity, contributions to local government coffers through taxes, as well as labour force demands. A business, in the context of the BRD, had one or more establishments or business units that independently reported statistics on revenues, expenses, and employment counts. Excluded from the BRD were small unincorporated businesses: those that were without corporate registration, had sales of less than $30,000, and no employee payroll. Establishments or units that reported to a business were the basic unit of observation in the BRD. It contained summary information on business employment in the county. It also provided a count of establishments in terms of the duration of employment, that being indeterminate, as well as eight class ranges based upon the number of employees. The user is cautioned not to use the BRD to generate employment estimates for industries since the data were unreliable for the task. The database had two levels of geographic resolution. Data were available at the sub-county level for seven Census Sub-Divisions (CSDs) (see Figure 5.2). Three CSDs correspond to the urban areas (CSD 1207004, Berwick; CSD 1207012, Kentville; CSD 1207024, Wolfville). The rural areas in the county from west to east were CSD 1207001, Kings Sub-Division A; CSD 1207016, Kings Sub-Division B; CSD 1207011, Kings Sub-Division C; and CSD 1207021, Kings Sub-Division D. All data at the CSD-level were summarized for the entire municipality (CD 1207) since the boundaries of CSDs fit by definition within the higher levels of Statistics Canada's hierarchical geographic classification system.

A third database recorded establishments primarily engaged in the manufacturing sector, including logging, summarized for the county via a custom request of the Annual Survey of Manufacturers and Loggers (ASML) from Statistics Canada. Note that the database was intended not to provide a statistical profile of manufacturing in the municipality but to contribute a representative, random sample of establishments in support of the ASML (Statistics Canada 2009b). The county was a major contributor to the province's food and non-food manufacturing sector, so the sample was sufficiently large to permit custom processing for this study. With respect to activities, establishments were categorized according to the North American Industrial Classification System (NAICS) on the basis of the manufacturing activity that contributed the largest portion of the value of sales. An establishment or business unit was defined as the smallest operating unit within a business that independently reported sales, expenses, inventories, and employee counts. The establishments were further divided into non-food and food manufacturing categories, with the latter further refined into the animal slaughtering and processing category in NAICS (in other words, comparable to first-order manufacturing, second-order

Learning about the Agricultural and Food System in Your Municipality | 101

livestock-based manufacturing in the classification scheme of this study). Activities associated with this class included "slaughtering animals, including poultry; preparing processed meats and meat by-products; or rendering animal fat, bones and meat scraps" (Statistics Canada 2007a). The geographic resolution of the database was at the scale of the county. It was

Figure 5.2. Map of Statistics Canada's Census Division 1207 and Sub-Divisions for Kings County.

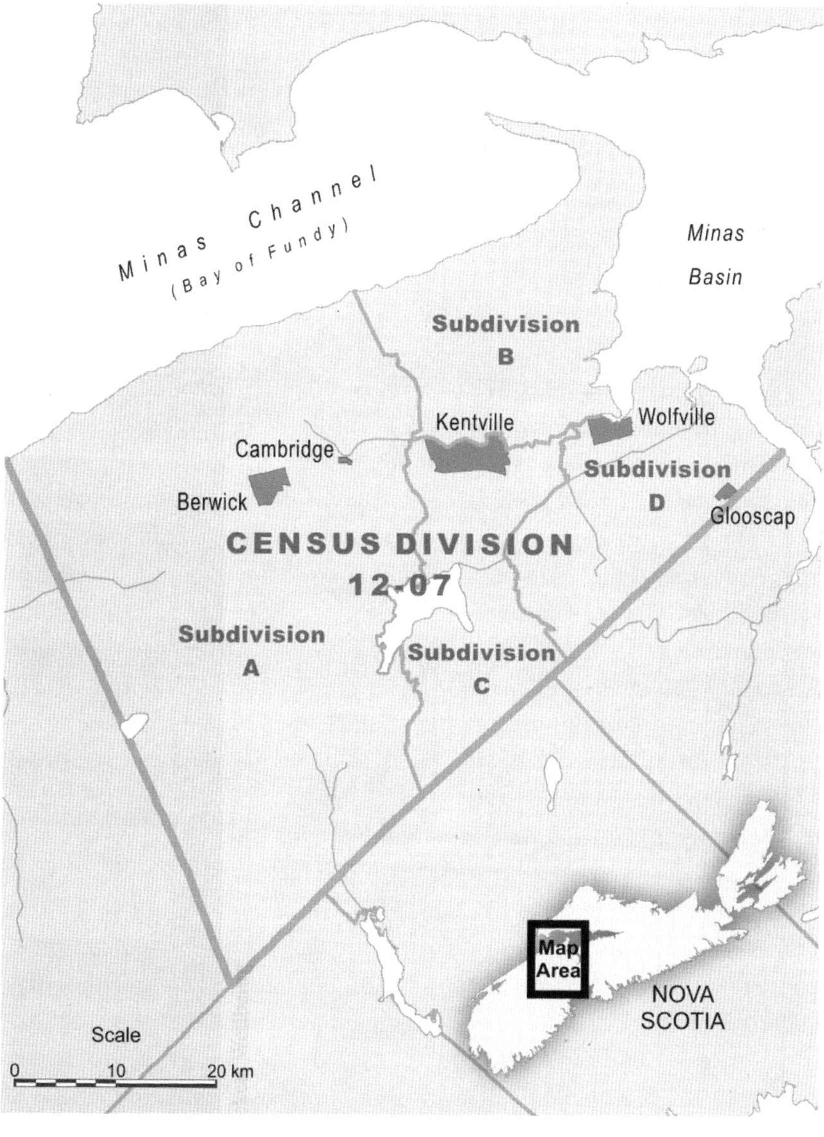

Source: Statistics Canada (2007b).

not possible to identify the geographic location of clusters or groups within the municipality. Nor was it possible to produce data for areas within the municipality to ensure the confidentiality of survey respondents.

The Kings Regional Development Agency's online Business Directory Database (BDD) was the fourth database. It provides businesses within the Annapolis Valley with a means to raise awareness of their activities among existing and new clients (Kings Regional Development Agency 2010). Since it was the business owner's decision whether or not to list, the BDD was not a statistical or representative sample of businesses in the valley. However, the database had hundreds of records; thus, it provided insight on business activities within the agricultural and food system. Permission was received from the agency to create a custom database using the activity classification scheme, hereafter noted as the BDD-CP. The geographic resolution of the database was at the scale of street address in the county. Although the BDD could not be mapped, Kings County planning staff assigned a street address and error-checked the records so that the BDD-CP could be geographically displayed and analyzed.

A recent food cost study in Nova Scotia provided the foundation to estimate the annual expenditures of residents of the county to acquire healthy diets. Using primary data collection and analysis of interviews with people and agencies concerned with income-related food insecurity, researchers from Mount Saint Vincent University determined provincial coefficients of the average weekly costs of nutritious diets (Mount Saint Vincent University 2009). The coefficients were refined by gender and age. To provide an estimate of the annual expenditure for a nutritious diet for people in the county, this study applied those dollar coefficients to Statistics Canada's 2006 Census of Population (Statistics Canada 2007b). The database was a proxy for the cost of a healthy diet for residents of the county and was completed for the entire municipality.

Based upon these five databases, results are considered in the following paragraphs, beginning with the primary production activity class. As of 2006, just over 600 farms were recorded in the custom *Census of Agriculture* profile for the municipality. These operations generated annual incomes and expenses exceeding $300 million in 2005. The total farm area was slightly more than 120,000 acres or about 20 percent of the total land base of the county. Total capital investment exceeded $400 million, with about 75 percent of that total invested in land and buildings. Livestock-based farming systems accounted for less than 20 percent of the total number of farms yet reported 55 percent of gross sales from agricultural production in the municipality. Of importance were chickens for broiler and egg production and cattle for dairy production. Although hog production was also important, members of the technical steering committee noted that severe financial difficulties had virtually eliminated this type of farming in the municipality at the time of the study. With regard to crop production, operations involved with vegetables were significant relative to the area of common field and forage crops. This type of farming accounted for less than 10 percent of all farms, yet it reported over 15 percent of total gross sales from agricultural production. In total, these four types of farming accounted for just one in four farms but three of four dollars of total gross sales.

Two other databases provided information on the primary production activity class, including the BRD and BDD-CP. In the case of the former, it was found that this class was concentrated in Kings A and B, with lower percentages in Berwick and Kings D (see Table 5.2).

With regard to the BDD-CP, over 130 records showed the manufacturing and retail trade activity classes in addition to the primary production activity class (see Table 5.3). A typical business in this category is a farm that packages its own produce and sells its products via farmers' markets or farm gate sales. Just over twenty other records reported primary production in addition to activities from other classes but not the manufacturing or retail trade activity class.

The support services activity class was concentrated in one rural and one urban CSD. From the BRD, it was found that Kings B and Kentville each recorded about 30 percent of the more than twenty establishments. A much lower percentage was reported for Kings A, Berwick, and Kings D. The BDD-CP reported almost thirty businesses in this class, with an additional five records showing the support services activity class with other activities.

Nearly forty establishments reported in the BRD were assigned to the manufacturing activity class. Two rural areas, Kings A and B, had more than half, with another almost 20 percent reported in Kentville and a lesser percentage reported in Kings D. According to the ASML, over $900 million of annual income and expenses was recorded by the twenty-four establishments involved with food manufacturing. In the Second-Order Activity Classes, the livestock-based manufacturing activity class (i.e., meat product manufacturing, more specifically the six establishments recorded to animal slaughtering and processing NAICS codes) accounted for nearly half of the total number of establishments. In terms of the crop-based manufacturing

Table 5.2. Number of establishments by First-Order Activity Classes from Statistics Canada's 2010 BRD for Kings County.

FIRST-ORDER ACTIVITY CLASS	NUMBER OF ESTABLISHMENTS	KINGS A	BERWICK	KINGS B	KENTVILLE	KINGS C	WOLFVILLE	KINGS D
PRIMARY PRODUCTION	402	27%	14%	33%	6%	2%	4%	14%
SUPPORT SERVICES TO PRIMARY PRODUCTION	21	14%	10%	33%	29%	0%	5%	10%
MANUFACTURING	38	21%	5%	34%	18%	5%	3%	13%
WHOLESALER AND DISTRIBUTOR TRADE	43	26%	14%	33%	19%	7%	2%	0%
RETAIL TRADE	95	33%	7%	15%	16%	18%	5%	6%
TRANSPORTATION AND WAREHOUSING	1	0%	0%	100%	0%	0%	0%	0%
NON-FOOD SERVICES (PROFESSIONAL/SCIENTIFIC/TECHNICAL)	0	0%	0%	0%	0%	0%	0%	0%
ACCOMMODATION AND FOOD SERVICES	137	19%	8%	11%	16%	21%	20%	5%

Source: Kings Regional Development Agency (2010).

Table 5.3. Number of businesses assigned to one or more activity classes in Kings County from the BDD-CP.

ACTIVITY CLASSES	NUMBER OF BUSINESSES REPORTING
PRIMARY PRODUCTION, MANUFACTURING, AND RETAIL TRADE	133
PRIMARY PRODUCTION AND ALL OTHER COMBINATIONS	22
SUPPORT SERVICES	29
SUPPORT SERVICES AND ALL OTHER COMBINATIONS	5
MANUFACTURING AND ALL OTHER COMBINATIONS EXCEPT PRIMARY PRODUCTION	28
WHOLESALER AND DISTRIBUTOR TRADE AND ALL OTHER COMBINATIONS EXCEPT PRIMARY PRODUCTION	7
RETAIL TRADE	118
RETAIL TRADE AND ALL OTHER COMBINATIONS EXCEPT PRIMARY PRODUCTION	6
TRANSPORTATION AND WAREHOUSING	10
SERVICES (PROFESSIONAL, TECHNICAL, SCIENTIFIC)	39
ACCOMMODATION AND FOOD SERVICES	163

Source: Statistics Canada (2007b).

Table 5.4. Summary of estimated annual expenditures for a nutritious diet in Kings County (2008 dollars).

AGE RANGE	2006 CENSUS OF POPULATION		2008 ADJUSTED ANNUAL EXPENDITURES	
	FEMALE	MALE	FEMALE	MALE
0 TO 19	7,230	7,330	$11,977,649	$14,668,865
20 TO 64	18,560	17,615	$33,223,367	$41,234,053
65 AND OVER	5,165	4,120	$8,892,754	$8,454,201
TOTAL	30,955	29,065	$54,093,770	$64,357,119

Source: Morton Horticultural Associates (2009). *A Value-Added Strategy for Agriculture: Kings County, Nova Scotia.* FY 2009–2014. Cold Brook, NS: Morton Horticultural Associates.

activity class, establishments in the NAICS other food manufacturing category were important, especially with large enterprises in frozen food manufacturing as well as fruit and vegetable canning, pickling, and drying NAICS codes. From the BDD-CP, it was found that nearly thirty establishments reported agriculture-related manufacturing activities but not the primary production activity class. As noted above, another 130 businesses reported primary production and retail trade activity classes as well as the manufacturing activity class.

With regard to the wholesaler and distributor trade activity class, the BRD showed for the slightly more than forty establishments that Kings A and B recorded almost 60 percent of that total. Lower percentages were recorded for Kentville and Berwick. The BDD-CP reported fewer than ten businesses in this class.

Two databases contained information that identified establishments in the retail trade activity class. The BRD reported ninety-five businesses, with one-third located in Kings A and another half equally distributed among Kings B, Kentville, and Kings C. The BDD-CP database showed nearly 120 businesses reporting only this activity, with an additional 130 businesses reporting primary production and manufacturing activity classes as well as the retail trade activity class.

It was possible to characterize the transportation and warehousing activity class and the non-food services (professional/scientific/technical) activity class in only a cursory way. The BRD reported one establishment in the former class and no records in the latter class. From the BDD-CP, ten businesses were reported in the former class and nearly forty in the latter class.

In terms of the accommodation and food services activity class, the BRD reported nearly 140 businesses, with four areas accounting for three-quarters of them in roughly similar proportion: Kings A, Kentville, Kings C, and

Wolfville. The BDD-CP database showed over 160 businesses that reported only this activity, typically restaurants or bed-and-breakfast operations.

No information was available in the databases to describe the waste management activity class.

With regard to the consumers activity class, it was estimated that total expenditures for a nutritious diet by the residents of the municipality approached $120 million each year (see Table 5.4). More than half of that total was related to the age group twenty to sixty-four, with an estimated value of almost $75 million. Note the much smaller population counts in the age group zero to nineteen, which will influence food-related expenditures as those aged twenty to sixty-four begin to move into the age group sixty-five and over. Without substantial in-migration, there will be a considerable drop in the number of people and expenditures in the twenty to sixty-four cohort in coming decades.

The results showed concentrations of activity in the agricultural and food system in various areas of the municipality. The predominantly rural Kings A reported a relatively large proportion of businesses in six activity classes (see Figure 5.3). Berwick, a smaller Census Sub-Division with a higher proportion of urban population, reported similar results in terms of primary production, support services, and wholesale and distributor trade activity classes. Kentville, the primary urban centre in Kings County, showed high concentrations of six activity classes; however, the primary production activity class is less evident compared with other areas. Kings B, a predominantly rural subdivision, reported concentrations of activity in production, manufacturing, distribution, and retail trade aspects of the agricultural and food system. Finally, Kings D is a rural area with concentrations of primary production, support services, and manufacturing activity classes.

DISCUSSION OF STUDY RESULTS

The method has produced results that might challenge conventional thinking about the agricultural and food system within the municipality. For example, the considerable activity in different sectors of the agricultural and food system located in both rural and urban areas might be an unexpected outcome. Rural areas reported not only the primary production activity class but also support services, manufacturing, wholesale and distributor trade, retail trade, and accommodation and food services activity classes. Clearly, a careful examination of which sectors are located in rural areas is critical to gain a full appreciation of the agricultural and food system in the municipality. Urban areas, on the other hand, reported similar classes as well as non-food services and consumers activity classes. Reporting of support services, manufacturing, and several trade-related activity classes in an urban area might also be an unanticipated result. However, urban areas are important locations for businesses to access labour as well as the physical infrastructure necessary for the manufacturing and efficient movement of finished agricultural and agriculturally based products. Urban areas can also house intellectual resources that support the agricultural and food system, as in the case of the federal Atlantic Food and Horticulture Research Centre in Kentville. To effectively plan for the agricultural and food system, the type and concentration of activities located in rural and urban areas should be fully explored.

Significant financial activity is generated by the agricultural and food system within the county each year, especially from the primary production, manufacturing, and consumers activity classes. Gross income and summary expenses in 2005 as reported to the *Census of Agriculture* exceeded $300 million. In the case of the manufacturing activity class, the figure was closer to $1 billion (2008 dollars), according to the ASML. The estimated annual expenditures for the consumers activity class could approach $120 million (2008 dollars), for a healthy diet, though it was not possible to determine if, in fact, residents spent their incomes in this fashion. Some of this estimated financial activity can include double-counting, as in the case of expenditures by consumers made directly to producers. Also, the total financial activity would probably be much higher if it were possible to profile all activity classes. Even with these caveats, the results show that the agricultural and food system is a major financial driver of the local economy on an annual basis.

The importance of the technical steering committee was evident in several aspects of the study. The committee provided guidance in the inventory of the sixty-five databases noted in the course of this study. Committee members contributed to discussions on the classification scheme and potential value of different classes to planning for the agricultural and food system. The committee provided local expertise for the interpretation of results, as noted in the case of the demise of hog production. Because, as discussed in the introduction, the agricultural and food system lacks a standard configuration beyond a conceptual level at local and regional levels, a technical steering committee should remain integral to the method.

SELECTED TOPICS FOR FURTHER CONSIDERATION

The topic of highest priority for further consideration is continued demand from the ultimate user of information products generated by the method: that is, the county council. Lower-tier governments have generally been exceptionally cautious when exercising their influence as key partners in development of the agricultural and food system. The degree to which local and regional municipalities embrace this role will have a profound bearing on the demand for future efforts to characterize more effectively the agricultural and food system. Kings County is a municipality that has demonstrated this resolve. Beginning in 2011, the county council announced a land-use planning process called Kings 2050. The principal deliverable is a strategy by which the municipality and its partners in coming decades can "balance the needs of a significant

Figure 5.3. Map showing concentrations of First-Order Activity Classes of the agricultural and food system by Census Sub-Division in Kings County.

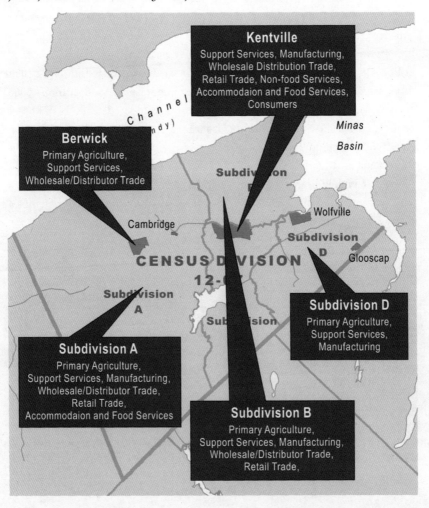

resource based economy, particularly related to agriculture, with a consistently growing non-resource based population and economy" (Kings County n.d.). Development of the strategy to support council's decision-making process might be the impetus to research existing databases and partnerships.

The demonstrated interest of council can be a motivation to reconsider the type and accessibility of data collected on the agricultural and food system in the county. For example, all businesses operating within the municipality must obtain a business licence each year. With minor adjustments to the form, a business owner in the agricultural and food system could quickly identify relevant First- and Second-Order Activity Classes. In this way, a spatial database of considerable value could be produced at minimal expense. Should the province wish to assist municipalities more actively, perhaps several databases under the control of this level of government could be summarized in a consistent manner while fully respecting privacy regulations arising from the legislation. Genuine efforts to grasp more fully and plan for enhanced viability of the agricultural and food system can also bring another important partner to the planning table: the associations that represent different groups throughout the system. Associations maintain valuable data on their members that would help to characterize more accurately their particular contributions to the system. Three of the five databases used in this study were acquired from the federal government in full respect of the confidentiality of respondents, and local municipalities might consider encouraging other organizations to pursue with them cooperative solutions to the development and exchange of topical information.

With regard to the method, at least four refinements could yield superior results. The first refinement is an accepted spatial base unit at the sub-municipal level that could document clusters of activity in the agricultural and food system and where they occur within the municipality. This study found that rural and urban areas reported different aspects of the system based upon boundaries defined by Statistics Canada. An accepted spatial framework for data collection and analysis would also permit the integration of other databases into a common base unit. More detailed, robust profiles of the agricultural and food system from multiple databases would support focused planning at the sub-municipal level.

A second refinement is better characterization of the social dimension of the agricultural and food system. For this project, expenditure coefficients from the Mount Saint Vincent University nutrition study were used to profile the consumers activity class. As male and female population counts by age are available from Statistics Canada for smaller areas, a comparable database could be generated to the level of census tracts (CTs). CTs are fairly stable delineations of small areas with a total population of 2,500 to 8,000 people. This database would be useful in explorations of links between concentrations of population and aspects of the agricultural and food system identified within a CT, such as the retail trade and food/beverage stores activity classes. Similar information has been used to explore food security issues such as physical access to a nutritious food supply.

Another important aspect of the social dimension is employment in the agricultural and food system. Labour force participation brings to the fore both socio-economic and socio-cultural topics of interest to land-use planners. One example is the rapid demise of hog production within the county immediately before the start of this project. Economic impacts were felt not only in the primary production activity class but also throughout the agricultural and food system as fewer hogs were raised, transported for processing, and distributed through wholesale and retail businesses. Also, farm families and the broader community had to manage additional stresses arising from this substantial disruption to the system. The ability of families and communities to cope is influenced by the availability of community and social support networks. However, no information on this vital aspect of social capital was located for inclusion in the non-food (professional, technical, and scientific) activity class. Characterization of these networks using measures of social capital can generate insights into the resiliency of the agricultural and food system.

Increased representation of the natural capital needed to sustain the agricultural and food system is a third refinement that should be addressed. No database was identified that represents links between First-Order and Second-Order Activity Classes and the physical as well as biological resources in the municipality. This information gap is disconcerting for at least two reasons. First, it is not possible to determine the ecological goods and services currently required to maintain the agricultural and food system as well as other valued land uses in the municipality. Second, it is a challenge to deliberate on the full costs and benefits of the system to the environment compared with other land uses in the municipality without this information. Inevitably, trade-offs among land uses are inherent in the current and future development of the county. In the absence of representation of the links within and between land uses and the natural resources required to sustain them, it is indeed a daunting task for a council to determine the fine balance among them needed to achieve sustainable development.

A fourth refinement is the systematic evaluation of the degree to which replicable results can be obtained by different users of the method. A vast amount of data can be collected on the agricultural and food system, as found in this study. However, each database has been created to meet a distinct purpose, often using different data collection methods and classification schemes. For a user to utilize these databases effectively to profile the agricultural and food system, each record must be consistently assigned to the relevant First-Order and Second-Order Activity Classes. That assignment requires assumptions to be made so that a record showing; for instance, a farm selling carrots is coded as primary production, less conventional crop activity, classes; manufacturing, crop-based manufacturing activity, classes; and retail trade, other retail trade, activity classes. The degree to which classification assumptions vary among users and therefore influence the consistency of the results should be evaluated. With several iterations of the assignment process in major agricultural regions of Canada, an explicit set of decision rules or protocols could be developed. With an agreed set of rules, it would then be feasible to automate the classification process to realize greater efficiencies in future applications of the method.

CONCLUSION

The prototype integrated existing data from several sources and produced a profile showing that the agricultural and food system was an important land use in the municipality. Although tantalizing information was gleaned on the production, manufacturing, and trade of certain crop- and livestock-based products, as well as the different combinations of activities within and between rural and urban areas, the complexity of the agricultural and food system is far greater than might have been previously appreciated. Application of the method to the considerable data collected on the system but not leveraged for this project presents an opportunity to generate a more comprehensive description in the short term. Representation of the social and environmental capital that sustains the agricultural and food system along with refinement of the classification protocol would generate new insights of value in the land-use planning process. The knowledge gaps noted in this study will be overcome if the municipal council continues to demand products that help it to determine the balance required between the agricultural and food system and other land uses so that Kings County remains on a path toward sustainable development.

REFERENCES

Born, H., and J. Bachmann. 2006. "Adding Value to Farm Products: An Overview." http://www.attra.ncat.org/attra-pub/valueovr.html.

British Columbia. Ministry of Agriculture and Lands. 2007a. *Public Amenity Benefits and Ecological Services Provided by Farmland to Local Communities in the Fraser Valley: A Case Study in Abbotsford, B.C.* Strengthening Farming Report File No. 800.100-1. Abbotsford: Ministry of Agriculture and Lands.

———. 2008b. *The Economic Impact of Agriculture in Abbotsford (Market Based Goods and Services).* Abbotsford: Ministry of Agriculture and Lands and Abbotsford Chamber of Commerce.

Canadian Co-Operative Association. 2008. "Opportunities in Value-Added Agriculture." http://www.coopscanada.coop/en/orphan/Opportunities-in-Value-Added-Agriculture.

———. 2008. "The Lay of the Land: Local Food Initiatives in Canada."

Devanney, M. 2006. *The Benefits of Agri-Food Value Chains.* Halifax, NS: Department of Agriculture/Aquaculture and Fisheries, Business Development and Economics.

Federation of Canadian Municipalities. 2012. *The State of Canada's Cities and Communities 2012.* Ottawa: Federation of Canadian Municipalities.

Food Secure Canada. 2011. "People's Food Policy Project." http://peoplesfoodpolicy.ca/home.

Harry Cummings and Associates. 1999. *Economic Impact of Agriculture on the Economy of Simcoe County.* Draft report to the Simcoe County Board of Education, Simcoe County Federation of Agriculture, and Human Resources Development Canada. Guelph: Harry Cummings and Associates.

Hiley, J.C., G. Bonneau, K. Thomas, and M. Rousseau. 2011. "Canadian Municipalities Satisfy the Craving for Sustainable Agriculture and Food Systems." *Municipal World: Canada's Municipal Magazine* 121 (4): 27–30.

Hofmann, N., G. Filoso, and M. Schofield. 2005. "The Loss of Dependable Agricultural Land in Canada." *Rural and Small Town Canada Analysis Bulletin* 6, 1. Statistics Canada Catalogue No. 21-006-XIE.

Kings County. N.d. "Municipal Planning Strategy." http://www.county.kings.ns.ca/residents/planning/mps.htm.

Kings Regional Development Agency. 2010. "Business Directory." http://www.kingsrda.ca/common/bdir.aspx.

Michiels, P.H., and J.C. Hiley. 2010. *Agricultural Profiling for Kings County, Nova Scotia: A Decision Support Tool for Agricultural Land Use Planning.* Brandon: Agriculture and Agri-Food Canada, Agri-Environment Services Branch.

Morton, G. 1995. "Value Added or Is that Added Value?" *Farm Business Management and Economics* 6 (3). Kentville: Nova Scotia Department of Agriculture and Marketing.

Morton Horticultural Associates. 2009. *A Value-Added Strategy for Agriculture: Kings County, Nova Scotia.* FY 2009–2014. Cold Brook, NS: Morton Horticultural Associates.

Mount Saint Vincent University. 2009. *Cost and Affordability of a Nutritious Diet in Nova Scotia: Report of 2008 Participatory Food Costing.* Halifax: Mount Saint Vincent University.

Reimer, B. 2003. "Understanding Social Capital: Its Nature and Manifestations in Rural Canada." http://nre.concordia.ca/ftprootFull/Inetpub%2002june2010/Inetpub/ftproot/ferintosh_papers/Bill_Reimer_SOCIAL_CAPITAL.PDF.

Smith, B., and S. Haid. 2004. "The Rural-Urban Connection: Growing Together in Greater Vancouver." *Plan Canada* 44 (1): 36–39.

Statistics Canada. 2006. *2006 Census of Agriculture Questionnaire.* http://www.statcan.gc.ca/ca-ra2006/q/q06-eng.pdf.

———. 2007a. "North American Industrial Classification System—31161—Animal Slaughtering and Processing." http://stds.statcan.gc.ca/naics-scian/2007/cs-rc-eng.asp?criteria=3116.

———. 2007b. "Kings, Nova Scotia (Code 1207)." Table. "2006 Community Profiles." 2006 Census. Statistics Canada Catalogue No. 92-591-XWE. http://www12.statcan.ca/census-recensement/2006/dp-pd/prof/92-591/index.cfm?Lang=E.

———. 2007c. "North American Industrial Classification System." http://www23.statcan.gc.ca/imdb/p3VD.pl?Function=getVDPage1&db=imdb&dis=2&adm=8&TVD=118464.

———. 2009a. *BRD-DRE: Definitions and Concepts Used in the Business Register.* Ottawa: Statistics Canada.

———. 2009b. *Reporting Guide: Annual Survey of Manufacturers and Logging, 2009.* Ottawa: Statistics Canada.

CHAPTER SIX

Smart Growth in Ontario: Getting Ahead of the Future

GARY DAVIDSON

During the summer of 2004, Ontario released several discussion papers aimed at reforming the planning system in that province and proposing large-area integrated plans. These papers were strengthened in 2005 by the Greenbelt Act and proposed amendments to the Planning Act and Conservation Land Act. The government intended to address several major planning problems that have gained public profiles in the past decade and provide specific plans for various high-growth regions in the province. These planning reforms and large-area integrated plans represented the most concerted effort to change planning directions in a generation and guided planning in Ontario toward a more compact and urbanized future. In general, they fall under the collective name of Smart Growth initiatives.

Although many planning issues such as sprawl and gridlock are seen primarily as urban problems, their impacts and solutions reside equally in rural and urban areas. Farmland preservation advocates have sought to address some of the myriad changes under way and to assess their impacts on rural and agricultural communities.

This chapter presents some reflections on the Smart Growth initiatives for rural and agricultural communities in Ontario. These are reflections from the vantage point of professional practice gained from working in rural communities for over forty years both in Canada and abroad. The chapter shares, often in abbreviated and summary form, some of the key aspects of the Smart Growth initiatives that rural and agricultural communities need to consider when planning for the future from a comprehensive and sustainable perspective.

Over the past forty years, there have been numerous efforts in Ontario directed at large-area integrated planning. These efforts have had several names: comprehensive planning, regional planning, rural community economic development, and the current offering, Smart Growth. All of these approaches have had profound impacts on rural and agricultural areas.

A major theme of these approaches in rural areas has always been the protection of agricultural land, no doubt a worthy cause. However, a singular focus on protecting agricultural land has not helped rural communities in the past and will not assist them in the future. The focus needs to shift from restricting land use to fostering the long-term sustainability of rural and agricultural communities. This might sound similar, but there is a chasm of difference. The agricultural community is an integrated combination of four critical elements:

1. the resource base (agricultural land in this case);
2. the economy;
3. the community; and
4. the environment.

Applying this fourfold context and not merely assessing the easier quantum of the protection of farmland is critical. Protecting farmland looks to the past, whereas fostering the long-term sustainability of the agricultural community exists in the domain of "getting ahead of the future." One is preservative and reactive; the other is strategic and proactive. This four-dimensional matrix is used to consider the numerous efforts, policies, legislative proposals, and papers circulated by the Province of Ontario.

SMART GROWTH QUESTIONS

The current Smart Growth initiatives in Ontario involve actions grouped under three general directions: planning reform, large-area integrated planning, and infrastructure renewal. All of them have impacts on rural and agricultural communities. This chapter does not attempt to provide a detailed review. Rather, those areas that affect rural regions in specific and significant ways are addressed.

There are two fundamental questions when considering the current Smart Growth range of initiatives.

The first question is about implementation. *Will today's Smart Growth efforts prove to be better than previous attempts at large-area integrative planning?* Over the past forty years, Ontario has launched numerous initiatives focused on such large-scale planning, including the Toronto Centered Region Plan in the early 1970s, the major regional planning and governmental reorganization plans of the 1980s, and the Provincial Policy Statement (PPS) directives of the 1990s. Now, in the early 2000s, we have Smart Growth. Under all of these aliases, integrated planning and development have been fostered. Yet all political parties have floated the same ideas, all of which stem from basic comprehensive planning theory. Still, considerable land-use problems remain unresolved. Cynics point to past failures while optimists express hope. It is not that black and white, of course, but the question that many people ask is why will this attempt be any different, or any more successful, than previous efforts? An assessment needs to be made of whether current Smart Growth initiatives will fare any better than past attempts with the same approaches. Or is Smart Growth really just a marketing tool?

The second question is about impact. *What will be the impact of Smart Growth policies and programs on rural and agricultural communities?* This is the central question. If the current programs, policies, and strategies are implemented, then will they help or hinder rural and agricultural communities? This is the key consideration for those concerned about the future of rural areas. Even if Smart Growth policies and programs are successfully implemented, it is not axiomatic that they will benefit rural communities or achieve long-term sustainability of agricultural areas.

SMART GROWTH AND THE AGRICULTURAL COMMUNITY

As noted, the agricultural community needs to be considered from four perspectives: land, economy, community, and natural environment. These four aspects provide the frame for an analysis of the impacts of current Smart Growth initiatives. The analysis focuses on the balance across these four dimensions. If a sustainable agricultural community is the goal, then the balance is critical. This is a comprehensive, sustainable, and long-term view of the agricultural community, but it is not the only view. There is also a lexical viewpoint that proposes a prioritization of the elements. This approach makes one perspective, normally protecting agricultural land or conserving the natural environment, the main priority to which all others must relate. In rural and agricultural communities, such a lexical approach is as confining as it would be if applied to urban communities, which it almost never is.

This chapter adopts a comprehensive perspective as the approach to planning in rural and agricultural communities; therefore, the balance among the four elements becomes critical. All of the initiatives must be assessed in terms of their balance across the four dimensions before any overall conclusion can be drawn. That Smart Growth might save agricultural land at the expense of the rural community is not acceptable. Assessing the qualitative nature of this balance across the four critical dimensions is complex and perplexing. However, the fact that it is difficult is no reason to shy away from it. Simple or, worse, simplistic solutions have created the problems that we now face.

A consideration of the Smart Growth documents, both singularly and collectively, provides an excellent vehicle to review the current state of Smart Growth and its impacts on agricultural and rural communities. Below, in point form, are some central aspects of the expected impacts, followed by some general observations and initial conclusions.

GENERAL OVERVIEW

Smart Growth initiatives, taken as a package and applied to the agricultural community, are primarily a series of specific land-use planning policies for the protection of agricultural land and the conservation and enhancement of the natural environment, supplemented by a variety of general community development goals. They represent a continuation of a lexical approach to planning in rural areas. Following are some important overall considerations that set the context.

- Smart Growth is primarily an urban strategy, and the relationship to the agricultural and rural community is inferential and confined to land-use directives at the level of implementation.

- Changing the key reference point in the PPS from "having regard to" to "be consistent with" actually restricts the ability of agricultural areas to plan for their sustainable futures.

- The plan for the Greater Golden Horseshoe is the first example of an implementation approach; although it focuses on agricultural land protection, it does provide for numerous opportunities to expand various urban uses into agricultural areas.

- The Greenbelt Plan, which covers large agricultural areas and numerous rural communities, places farmers in a park that makes future farm business planning problematic.

- Removing the Ontario Municipal Board from decisions on establishing and altering urban boundaries is one of the more positive aspects of the package for rural communities in the long term, even though altering urban boundaries is primarily an urban decision.

LOOKING THROUGH THE FOUR-DIMENSIONAL LENS

It is now instructive to look at the Smart Growth initiatives through the lens of each dimension that affects sustainability in agricultural and rural communities. In this way, an overall assessment of the balance among the elements can be advanced. And, through consideration of this balance, the probable impacts of the Smart Growth initiatives on the agricultural community can be determined.

1. *Agricultural Land*
    a. Protecting agricultural land seems to be the central Smart Growth theme when it comes to its rural strategy. Here the policies are detailed and prescriptive.
    b. The policies focus on three strategic directions: urban intensification, containment of urban settlement boundaries, and prohibitions on farmland conversion. Although positive, there are some dilemmas and caveats. For example, intensification of urban lands for commercial and industrial use is not addressed, and gravel extraction still holds the trump card in the rural land conversion game.
    c. Agricultural land protection policies are detailed and prescriptive. For example, the conversion of specialty crop lands such as tender fruit lands and the Holland Marsh is prohibited (Government of Ontario 2005, 37), agriculturally related severances are severely restricted (Government of Ontario 2005, 17), and a prohibition to enlarge settlement areas except during the comprehensive review of an official plan or based upon a comprehensive urban settlement boundary study is required.
    d. It is generally accepted that such policies are effective where they are applied over the long term to protect rural and agricultural lands. Several counties have long histories of such policies, going back to the 1970s. The real test will be whether the province will intervene in rural areas that do allow for urban development on prime agricultural lands.
    e. When these policies are operationalized, as in "A Growth Plan for the Greater Golden Horseshoe" (GGH Plan) through a series of priority urban centres and agricultural

land-use policies, various types of urban growth within rural areas outside existing urban boundaries are still permitted (Government of Ontario 2006, 16, 20–21).

However, in general, the Smart Growth suite of papers and plans tends to be very proactive in protecting the agricultural land resource, through policies to promote urban intensification, to stabilize urban settlement boundaries, and to restrict urban development on rural lands.

2. *Rural Economy*

   a. This is a perplexing area. The strategy documents all talk about promoting healthy, vibrant, and sustainable rural economies but offer little guidance, policy framework, or implementation detail to indicate how this will be accomplished.

   b. A key conceptual problem can be seen in the report from the Smart Growth Panel for Western Ontario. Although the vision is that "the agricultural sector and associated industries will continue to be promoted and enhanced," the active policy is protection of agricultural land (Western Ontario Smart Growth Panel 2003, 9, 25). It is well known that protecting agricultural land, while important, does not enhance or stimulate the rural economy. Without a strong rural economy, there is no sustainable rural community.

   c. The PPS is particularly strict about the ability of agricultural, industrial, commercial, and institutional activities to utilize rural space (Government of Ontario 2005, 27, 35). This has long been an issue in rural communities with strong agricultural land protection policies, for it blocks the required agriculturally related commercial and industrial infrastructure that supports agriculture.

   d. There is a lack of focus on how to achieve a strong rural economy. The goals contain appropriate sentiments, but the policies for implementation are particularly vague (Government of Ontario 2006, 25).

   e. There are too many references to "developing a strategy" and "establishing processes" and too little talk of details for implementation and even less for concrete action. This is in stark contrast to the urban components of the Smart Growth initiatives in which considerable details are presented.

The rural economy is a weak component of the Smart Growth initiatives for the rural community, both for agricultural areas and for small rural towns. There are many lofty vision statements and goals but few actual implementation policies or programs to support the goals. This is not surprising, for encouraging rural community economic development has been problematic for over a generation. Also, most of the detailed policies, in their quest to protect agricultural land, place further restrictions on sustainable rural development.

3. *Rural Community*
   a. Fostering the sustainability of the community, especially its social structures and institutions, is an important component of rural planning.
   b. The Smart Growth initiatives are probably the weakest in this dimension and retreat to generalities, often banal ones at that. They tend to treat the rural community as a sub-component of the agricultural economy (Western Ontario Smart Growth Panel 2003, 25–28).
   c. After two decades of removing educational, health, and cultural services from rural communities, discussion of developing long-term strategies, creating partnerships, and so on comes to sound hollow (Western Ontario Smart Growth Panel 2003, 35).
   d. The same theme resonates through educational services, well-being, and safety.
   e. Approaches to rural youth are similar: recognize the problem and suggest a strategy.
   f. A legitimate question is what will really change, and why haven't the rural approach and implementation policies been as well thought out as the urban ones? This portends a major concern since the Smart Growth initiatives claim to represent future directions for both urban and rural communities.

Like analysis of the rural economy, when the rural community is considered, details of implementation within the Smart Growth initiatives quickly fade. Most of the material under the dimension of the rural community that provides the framework for this analysis consists of general statements that verge on "wouldn't-it-be-nice-if" sentiments. Unfortunately, general goals and well wishes will not help rural communities to renew their lost and declining social infrastructure.

4. *Natural Environment*
   a. In this area, the policies are clearer and return to a more regulatory stance.
   b. Goals to protect and enhance the environment with respect to agricultural lands, water, and natural heritage are well structured, and implementation policies are developed in considerable detail (Government of Ontario 2005, 25–26).
   c. Protection of the natural environment, both functionally and aesthetically, helps both urban and rural communities. However, the burden of this protection falls more heavily on rural community residents than on urban community residents. There is a need to develop a mechanism to distribute these costs more equitably. Unfortunately, such an innovative policy approach is not even mentioned.
   d. Policies for protection and enhancement of the natural environment are the clearest of all the four dimensions analyzed in this chapter. It is a policy framework almost completely supportive of rural and agricultural communities.

A strong natural environment is critical to the functioning of a sustainable agricultural and rural community. The Smart Growth initiatives, both policy documents and plans, are strongly supportive of the protection and enhancement of the natural environment.

GENERAL OBSERVATIONS AND CONCLUSIONS

This chapter has provided an overview of the rather daunting task of considering the wide range of current Smart Growth initiatives and assessing their impacts on rural and agricultural areas. Since some of the reports, discussion papers, and laws have been released only recently, only general observations and conclusions can be offered. It will take decades to measure the impacts of these Smart Growth initiatives on rural communities. However, some general observations can be proposed.

- Smart Growth is primarily an urban growth and development policy, not a rural one. Although there are numerous benefits to rural areas from appropriate urban planning strategies, Smart Growth cannot be seen as a rural development strategy.

- There are some restrictions on the rural economy and a lack of specifics on community viability. In these areas, Smart Growth falls back to general goal statements and does not offer specific approaches to implementation.

- From a rural perspective, while protecting agricultural lands and the natural environment is an important consideration, the primary beneficiary is the urban community. No mechanism has been developed to help equalize the cost of this protection among rural and urban communities.

- The concept of strong rural communities is tied too much to protecting agricultural lands and the natural environment. Both are important in themselves, but they do not constitute a recipe for strong rural communities.

- Rural communities will not get much help from Smart Growth in their quest to rejuvenate and create sustainable, balanced communities. They will have to develop their own methods. This is consistent with most community development literature.

- Smart Growth is better at controlling and regulating urban growth than at promoting and enhancing rural communities. This has always been the problem with large-area integrated planning and regional planning theories when they are applied to rural areas.

At this early juncture, some key directions, and possibly conclusions, can be seen. They are preliminary but might help to provide a framework for further consideration and analysis of the Smart Growth initiatives in Ontario with respect to their impacts on rural and agricultural communities. Although recent legislative actions are encouraging, the question of whether current Smart Growth initiatives will be implemented over time will have to be revisited at a much later date when more information is available on both provincial and municipal planning activities and implementation programs.

If a comprehensive approach is the measure through which to review Smart Growth, then it fails, at present, the test of balance when applied to the rural community. Smart Growth policies and plans focus on protecting agricultural land and conserving the natural environment. In these dimensions, they are powerful and helpful to rural communities. However, in the dimensions of economy and community, there is a substantial weakness. In some ways, current Smart Growth policies will hamper the agricultural support economy from developing. With respect to community social structure, Smart Growth retreats to general goals, with no supporting implementation measures or dedicated funding. Rural communities will have to continue their quest on their own for strong, sustainable futures since Smart Growth and its rural and community planning strategies do not provide a sustainable framework.

Smart Growth needs to be seen primarily as an urban strategy. Rural areas can enjoy substantial benefits from effective management of urban growth. It is one of the major tools that rural areas need to undertake planning in their communities. If effective in the long term, then Smart Growth can bring some much-needed stability to the rural and agricultural land market that would benefit primary producers.

Finally, rural communities need to work within Smart Growth approaches to support urban intensification policies and attempt to broaden the policy framework within which rural communities plan. This should be aimed first at achieving a more favourable balance among land, environment, economy, and community and second at trying to balance the inherent costs between urban and rural communities that flow from land-use protection policies as they are applied to rural areas. This is a critical strategy for rural communities since, for some time to come, current Smart Growth policies will set the framework for planning in Ontario.

REFERENCES

Government of Ontario. 2006. *Growth Plan for the Greater Golden Horseshoe*. https://www.placestogrow.ca/content/ggh/2013-06-10-Growth-Plan-for-the-GGH-EN.pdf.

Government of Ontario. 2005. *Provincial Policy Statement, 2005*. http://www.mah.gov.on.ca/Page1485.aspx.

Western Ontario Smart Growth Panel. 2003. *Shape the Future: Western Ontario Smart Growth Panel Final Report*. Toronto, ON: Smart Growth Secretariat.

CHAPTER SEVEN

Farmland Protection and Livable Communities in British Columbia

KEVIN MCNANEY AND KELSEY LANG

The rapid growth of Canada's cities and towns over the past sixty years has presented some real challenges and lessons for planning for the future. There seems to be an emerging recognition that the growth pattern embraced by communities across the country is no longer working. This pattern is characterized by dispersed residential tracts, big-box retail "power centres," and places where the automobile is the only choice for transportation. Even more daunting is the relentless loss of farmland and open space that comes with this constant, outward, urban expansion.

Often forgotten is that this pattern of outward expansion, and the resulting urban form, are not unanticipated outcomes of the free market but conscious planning decisions in the postwar period. What we now call "urban sprawl" is the product of a highly sophisticated and self-reinforcing system designed to produce a uniform product and growth pattern. In planning circles, we often talk about the need to integrate transportation and land use, subsidies and regulations, and financial incentives and disincentives to achieve our planning goals. In the postwar period, we did just that by integrating insurance, building standards, development financing, engineering standards, transportation investment, land-use planning, and social planning. We created the system, and we got just what we asked for: urban sprawl.

Although this form of urban development did meet our needs for housing the baby boom and building the demand for consumer goods, the same demographic and financial pressures are telling us that this pattern needs to be reconsidered. Current issues that spark the demand for new patterns of urban growth include the following:

- Families are much smaller, and people are living longer, thus creating a need for more diverse forms of housing close to shops and services.

- People are growing tired of long commutes on congested roads and are seeking affordable housing close to jobs and good jobs close to home.

- Taxes continue to rise as we build more and more expensive infrastructure (roads, sewers, water mains) in areas with fewer and fewer taxpayers.

- Housing continues to be an enormous problem for those who cannot afford to buy into the "one size fits all" approach of conventional, single-family developments.

- Finally, the environment continues to deteriorate as we turn productive farmland and environmentally sensitive areas into asphalt, and the quality of our air and water continues to diminish from the enormous pressures of a growing urban population.

Increasingly, communities are looking to the Smart Growth movement for solutions that achieve a higher quality of life. Smart Growth is comprised of urban development strategies to reduce sprawl that are fiscally, environmentally, and socially responsible. Smart Growth is development that enhances our quality of life, protects our environment, and uses tax revenues wisely. The principles of Smart Growth, as outlined in Figure 7.1, encourage new growth in existing urban areas rather than on the fringes of our cities and towns. This helps to maximize our investment in taxpayer-funded infrastructure and uses growth to provide new amenities and services in our communities while simultaneously protecting our precious farmlands and open spaces. As such, Smart Growth is not anti-growth but better defined as the "how" and "where" of growth.

As anyone who has followed planning theory throughout the years knows, there is nothing controversial about the principles of Smart Growth. Most official community plans across Canada contain all of these planning and policy objectives. What most communities seem to misunderstand, however, is that all of the best intentions for the Smart Growth principles of urban intensification, mixed use, and walkable neighbourhoods will never occur without strong policies for rural and farmland protection. Achieving Smart Growth objectives without containing urban growth through farmland protection is akin to trying to fill a swimming pool without walls.

Table 7.1. Principles of Smart Growth.

PRINCIPLES OF SMART GROWTH	
• AVOID URBAN SPRAWL BY PROMOTING COMPACT HUMAN SETTLEMENT THAT AVOIDS UNPLANNED GROWTH AND ENSURES EFFICIENT DEVELOPMENT. • MINIMIZE THE USE OF CARS BY ENCOURAGING WALKING, BICYCLING, AND PUBLIC TRANSIT. • PROTECT THE ECOLOGICAL INTEGRITY OF URBAN AND SUBURBAN AREAS. • MAINTAIN THE INTEGRITY OF A SECURE AND PRODUCTIVE AGRICULTURAL LAND BASE.	• PROMOTE ADEQUATE AND AFFORDABLE HOUSING. • PRESERVE, CREATE, AND LINK URBAN AND RURAL OPEN SPACES. • PROMOTE INNOVATIVE URBAN DEVELOPMENT THROUGH MIXED-USE AND ALTERNATIVE DEVELOPMENT STANDARDS. • ENSURE AN EARLY AND ONGOING ROLE FOR CITIZENS IN PLANNING, DESIGN, AND DEVELOPMENT PROCESSES.

Source: Northwest Environment Watch (2003). *Sprawl and Smart Growth in Greater Vancouver*. Report available at www.smartgrowth.bc.ca.

HAS BRITISH COLUMBIA FOUND THE ANSWER?

British Columbia is famous for its soaring mountains, dramatic coastline, and mild maritime climate. These attributes have attracted people from across Canada and around the world. Less known about British Columbia, though, are the pockets of highly productive agricultural land that line the southern valleys of the province. Forty years ago the need to house a growing population in expanding urban areas and the strong desire to protect these precious agricultural lands came to a conflicting and enlightening head, and the progressive steps taken by the citizens and government of British Columbia are worth exploring.

A productive, secure, agricultural land base in the province is vital to our ability to maintain agriculture as a viable industry, to secure our food supply, and to act as an urban containment boundary. A key economic driver, agriculture supports the livelihoods of over 330,000 British Columbians through agriculture, aquaculture, food processing, food wholesaling, and food retail and service industries (Enchin 2013). It does this while contributing over $40 billion to the provincial economy and providing for 48 percent of our food needs. Remarkably, this agricultural productivity occurs on less than 3 percent of the provincial land base. The continued success of agriculture in the province requires that non-farm uses be limited in agricultural areas. If not, as shown in other jurisdictions, urban-rural conflict can intensify to the point where farming simply is not viable.

The threat of urban encroachment on agricultural land in British Columbia is real. About 88 percent of British Columbians live in urban areas, with the fastest-growing urban areas located adjacent to the best agricultural lands (in the Lower Mainland and the Okanagan Valley) (WorkBC 2013). Prior to introduction of the provincial Agricultural Land Reserve (ALR) in 1972–74, British Columbia was losing 6,000 hectares of prime agricultural land each year to non-farm uses. Despite boundary changes since that time, the ALR remains approximately the same size as it was at conception (4.5 million hectares). In comparison, Washington State lost an average of 30,000 hectares per year to non-farm uses between 1982 and 1992 (Northwest Environment Watch 2003). Before the recent introduction of greenbelt legislation, Ontario was losing one square kilometre of prime agricultural land every day to bulldozers (Sierra Club of Canada 2003). The ALR is now lauded around the world as one of the most progressive pieces of legislation for land-use planning and is the envy of virtually all jurisdictions across North America.

The fact that most of the population of British Columbia lives and works in urban areas poses a constant threat of loss of the ALR to urban uses. There is strong demand for large single-family homes in suburban areas, and there is limited understanding among the public of the need to view agricultural land as more than a greenfield site waiting for a better use.

In spite of these challenges, the ALR has illustrated its success both in the preservation of agricultural land and in the containment of urban sprawl in favour of more compact communities. Smart Growth BC and Northwest Environment Watch jointly released a study that compared the growth patterns of Seattle and Vancouver during the 1990s and illustrated that Vancouver had significantly greater success in containing sprawl and creating compact, complete communities. The report stresses the role of the ALR as a powerful tool for urban containment: "If Vancouver had grown like Seattle (in the 1990s), data suggest it would have converted approximately 18,000 additional acres—an area equivalent to about one-eighth of the ALR within Greater Vancouver, or to about four-fifths the size of the city of Burnaby—to sprawling suburban development" (Northwest Environment Watch 2003, n.p.).

During the 1990s, Vancouver put almost 100 percent of its new metropolitan population growth into walkable neighbourhoods (defined as a population density greater than twelve residents per acre). In comparison, Seattle put 55 percent of its new growth into car-dependent, sprawling neighbourhoods. Although the two cities experienced similar rates of population and economic growth during that time, a principal reason for Vancouver's success in containing urban sprawl was the ALR. The region simply could not sprawl onto some of the best farmland in the province, and the cities of the Lower Mainland were forced to direct growth into compact, walkable communities.

The effect of the ALR's urban containment characteristics on urban development across the province has been astounding. Large and small communities across British Columbia are continually lauded as being some of the most livable in the world. Although it is difficult to calculate a direct numerical relationship between the amount of growth and redevelopment in the province's central cities and the protection of farmland, there is little doubt that by protecting farmland the ALR has deflected this growth into the existing urban footprint.

Vancouver provides an excellent example of this phenomenon. The protection of farmland in the Fraser Valley has channelled residential growth into the central city, revitalizing communities and urban life. Whereas other downtown areas across North America have become hollowed out and deserted, the vibrancy of street life in Vancouver is similar to that of much larger urban centres. Planned communities have sprung up on old industrial "brownfield" sites such as the Coal Harbour, False Creek, and Yaletown districts. These areas provide housing for a mix of young urbanites and families. In fact, the demand for school sites in the downtown core has greatly exceeded the expectations of the Vancouver School Board, with seventy-five elementary schools and the announcement in 2012 that two more would be built (Vancouver School Board 2013). Although the suburb has long been the domain of the family, these new developments are a testament to the success of providing an environment conducive not only to young urbanites but also to children.

One of the most remarkable outcomes of this growth in central Vancouver rather than on the suburban fringe has been in transportation. Anyone who has ever tried to catch a bus in suburban areas is well aware of the difficulty of providing any transportation option other than the automobile in low-density areas. Public transportation and the ability to walk to services are simply not viable at population densities lower than approximately ten dwelling units per acre. The shift in transportation patterns in the downtown core of Vancouver after this explosive residential growth, however, astounded even city planners. From 1994 to 1999, the number of daily trips by automobile declined 13 percent from 116,198 to 101,371. Conversely, the number of daily trips by foot increased from 70,000 to 108,500—a jump of 55 percent (City of Vancouver 2002). And, as of 2008, approximately 40 percent of all trips in the city were conducted by transit, on bicycle, or on foot, with the goal of over 50 percent of trips being conducted this way by 2020 (City of Vancouver 2012).

While Vancouver demonstrates these innovative forms of urban growth and intensification on a large scale, this process is also evident in smaller communities across the province. Such communities are particularly susceptible to creating the vast subdivisions and strip malls associated with urban sprawl as they often lack full-time planning staff and are generally starved for any form of economic development to enhance the tax base. That many of these communities in British Columbia are surrounded by the ALR, however, has forced them to be efficient with land use. In fact, many of these communities are embracing innovative forms of urban intensification and development that one would expect only in larger centres. Some examples follow:

- In the Village of Pemberton, there is a mixed-use, multi-storey development that houses a grocery store, offices, and additional retail outlets in a compact form with shared parking.

- The Town of Oliver, in the Okanagan Valley, has branded itself as the "Wine Capital of Canada" and actively encourages the redevelopment of its downtown into a mixed-use, pedestrian-oriented village so as to protect the extensive vineyards and soft fruit orchards that generate wealth in the community.

- The City of Courtney on Vancouver Island has created the live-work residential development of Tin Town.

- An old hospital has been redeveloped using "green building" principles into the Vancouver Island Tech Park in the District of Saanich (Patterson 2004).

There are countless other examples of how communities across British Columbia have shown innovation and leadership in using land efficiently as a result of the inability to expand into the ALR.

In essence, the ALR has created boundaries obliging local governments to seek more innovative approaches to growth through densification rather than simply sprawling onto farmlands. The benefits from such approaches in terms of reduced automobile use and lower infrastructure costs are significant from both an economic perspective and an environmental perspective. Moreover, the proximity of an ALR radically increases the quality of life of urbanites, who benefit from its open spaces and important green infrastructure functions. Hence, protecting agricultural land through the ALR is a key strategy not only for protecting farmland but also for achieving Smart Growth objectives on the urban side of the boundary.

CHALLENGES TO THE ALR

Although in the past many local governments in British Columbia failed to see the importance of maintaining an agricultural land base, the trend is shifting toward conservation as more and more municipalities and regions complete agricultural plans. Still, these local governments are under immense pressures to provide services, infrastructure, housing, and local employment for their citizens, and constant outward expansion is seen as a key component of their economic strategies. Residential taxes help to support some of the financial burden of a local government, but the development of new industrial and commercial land is seen as the principal way to achieve fiscal viability at the local level. However, study after study has illustrated that converting agricultural lands to other uses often costs more in services than it produces in municipal tax revenues (American Farmland Trust 2003). Although many official community plans support agriculture, the lure of a development proposal is still a temptation that challenges that principle.

There was increased concern about ALR exclusions during the boom time of 2007 before the 2008 financial crisis hit, but this concern is less prominent today. In 2010, only 364 hectares were removed, the lowest annual amount since the ALR came into being. Although 632 hectares were removed in 2011, this was still the fourth-lowest annual amount and included only three hectares in the Lower Mainland. This, combined with some of the lowest levels of exclusion applications in years, shows that British Columbians widely accept and support the ALR (Mickleburgh 2012).

The rules of the ALR and its governing Agricultural Land Commission (ALC), which adjudicates exclusion applications, were changed on 1 November 2002, when the Agricultural Land Commission Act repealed the Agricultural Land Reserve Act, the Land Reserve Commission Act, and the Soil Conservation Act and replaced them with a new act. Key to these changes were the decisions to decentralize the commission by creating six regional panels of three commissioners each and to encourage the commission to devolve more authority to local governments on issues of land use within the reserve. Before this change, and since the 1970s, the ALR had successfully mitigated the constant threat of incremental urban encroachment onto agricultural land by maintaining decision making at the provincial level rather than the local level. The net effect is that the provincial "common good" of long-term agricultural security was paramount over the local desire for increased urban expansion. This is in direct contrast to virtually all other jurisdictions in North America, in which decisions on agricultural land use have been made at the local level and thus have been much more susceptible to pressure for development. With this regional panel structure, the ALC needs to remain vigilant to ensure that decision making is not influenced by local actors.

In 2003, the ALC added the goal of creating "a provincial land reserve system that considers community interests" in its vision. To achieve this goal, the ALC "will be working to balance the provincial interest in preserving lands for agriculture with community interests and the need for land for housing, employment, and community purposes" (BC Land Commission 2003). The concern is not about considering community interests per se but that these interests must be explicitly weighed against the primary and paramount goal of maintaining the integrity of the ALR.

There has also been a long-standing concern about the need for greater transparency in the assessment and decision-making processes concerning agricultural land and the activities of the ALC. Since 2005, the ALC has taken significant steps to increase this transparency, since it is important that all British Columbians understand the processes of the ALC and its regional panels. The public and media need to be able to monitor those deliberations and, if they desire, make comments. While respecting legitimate issues of confidentiality, reports that recommend or comment on applications to the ALC should be public documents, and any exchanges between applicants and commission members or staff should be divulged before decisions are made. This transparency will allow the public to monitor any changes to the ALR and anticipate any threats that might arise. This process will help to bring the notion of agricultural land preservation away from the simple processing of ALR applications and toward a provincial appreciation of and commitment to the integrity of the land reserve.

LIVABLE COMMUNITIES REQUIRE PRODUCTIVE FARMLAND
British Columbia has four decades of evidence demonstrating that policies for farmland protection have enormous benefits both for the maintenance of a productive farming industry and for the creation of compact, complete, and livable communities. Communities of all sizes across the province have shown that population and economic growth do not require constant urban expansion into farmland. In fact, the BC experience seems to indicate that a high degree of livability can be achieved successfully through farmland protection and urban intensification.

Arable land is the foundation of the agricultural industry, fundamental to our food security, and a critical buffer against uncontrolled urban expansion. We need to protect its future, and British Columbia has led the way with its innovative ALR. Jurisdictions across Canada can look west to see how a long-term vision has resulted in both a thriving agricultural industry and communities consistently rated as the most livable in the world.

ACKNOWLEDGEMENTS
Many of the ideas and much of the insight contained in this chapter are the collective work of the staff, board, and advisory committees of Smart Growth BC. I am very grateful for all of the insights into growth management and farmland protection that they have imparted to me.

REFERENCES

American Farmland Trust. 2003. "Cost of Community Services Studies." http://www.farmland.org.

BC Agricultural Land Commission. 2003. "Reserve Opinion."

City of Vancouver. 2002. "Downtown Transportation Plan." http://www.city.vancouver.bc.ca/dtp/.

——. 2012. *Transportation Plan 2040*. Vancouver: City of Vancouver.

Enchin, H. 2013. "Agriculture Is the All-Important Afterthought in B.C.'s Economy." *Vancouver Sun*, 26 January, J1, J3.

Mickleburgh, R. 2012. "B.C. Farmers' Anger Turns to Support for Agricultural Land Reserve." *Globe and Mail*, 14 September. http://www.theglobeandmail.com/news/british-columbia/bc-farmers-anger-turns-to-support-for-agricultural-land-reserve/article4547579/.

Northwest Environment Watch. 2003. "Sprawl and Smart Growth in Greater Vancouver." http://www.smartgrowth.bc.ca.

Patterson, G. 2004. "Agriculture and Urban Areas: The Benefits of Efficient Land Use." Smart Growth BC.

Sierra Club of Canada. 2003. "Sprawl Hurts Us All." http://www.sierraclub.ca.

Vancouver School Board. 2013. "Schools." http://www.vsb.bc.ca/schools.

WorkBC. 2013. "Regional Statistics." http://www.workbc.ca/Statistics/Regional-Profiles/Pages/Regional-Profiles.aspx.

CHAPTER EIGHT

Rural Non-Farm Development and the Agricultural Industry in Ontario

WAYNE CALDWELL, ARTHUR CHURCHYARD, AND CLAIRE DODDS

Ontario's agricultural industry has become the most intensive and diversified in Canada. At the same time, thousands of new lots for farm and residential purposes are fragmenting the landscape, introducing limits on agricultural development. The net costs and benefits of creating new lots comprise a long-standing academic and policy debate, ranging from early discussions of the impacts of non-farm development (see, e.g., Bryant and Russwurm 1979) to more recent methods of estimating financial costs of scattered non-farm lot creation (Red Deer County 2006).

In Ontario, lot creation is part of the land-use planning and development policy process, which must be consistent with the Provincial Policy Statement (PPS). The PPS lot creation policies have changed twice since 1990: once in 1996 and again in 2005. In both cases, the PPS further restricted the type of new lots permitted in prime agricultural areas.

To determine the effectiveness of the 1996 and 2005 policy changes, this research analyzes the number and type of lots created per year from 1990 to 2009 in the agricultural designations of 102 municipalities in Ontario. These municipalities contain virtually all of the prime agricultural areas in the province, with the exception of pockets in northern Ontario. After data were collected, each municipality received a summary of the number and type of new lots created in farm areas to verify. Surveys and interviews were also conducted to better understand local factors in lot creation trends. Detailed analyses were conducted at the provincial, regional, and upper-/single-tier levels across the province.

Findings suggest that the provincially led planning policy approach has increased effectiveness in reducing the rate of scattered residential lot creation in most agricultural designations across the province. The rate has decreased in agricultural designations at almost twice the rate of decrease in non-agricultural designations. Both the 1996 and the 2005 policy statements were followed by 48 percent and 59 percent decreases in residential lot creation, respectively (based upon the average number of lots created per year in each policy period). This does not mean that all agricultural areas have been protected sufficiently. Municipalities and landowners continue to pursue non-farm rural severances, adding to the cumulative impacts of non-farm residential development.

These findings, taken broadly, suggest that province-wide planning policies can reduce lot creation rates. Despite decreased rates in recent years, cumulative impacts of residential lot creation continue to threaten agricultural viability. These results are directly relevant to provincial policy reviews, such as current and future reviews of the Provincial Policy Statement, the 2015 Greenbelt Plan, and the ongoing creation of regional Ontario growth plans.

RESEARCH OBJECTIVES

This research fills an information gap created when severance applications were no longer required to be circulated to various ministries, a change introduced at the same time as the Planning Act in 1990. Before 1990, municipalities were required to circulate rural and agricultural severance applications to the Ontario Ministry of Agriculture, Food, and Rural Affairs (OMAFRA). The data were useful in informing provincial and local land-use policy and implementation. After introduction of the 1990 Planning Act, the majority of severance applications were no longer circulated to OMAFRA. To fill this information gap, this research collected and analyzed land division data for all severances in designated agricultural areas from 1990 to 2009. This information was collected in two phases, the first from 1990 to 2000, the second from 2000 to 2009. Files for 2000 were collected in both phases and compared to ensure that collection methods returned similar results, even as data storage methods changed from the first phase to the second.

This chapter discusses the following research objectives:

1. to document the number and purpose of lots created within rural and agricultural Ontario;

2. to determine the relationship between current provincial policy and the creation of rural non-farm lots; and

3. to provide quality information to assist with reviews of the PPS, the Greenbelt Plan, and provincial and municipal plans.

The chapter does not discuss in depth the following topics, which will be addressed further in future published analyses:

1. impacts of local land-use policy in effect when these lots were created;

2. economic, social, and environmental impacts on municipalities and the agricultural industry, as identified in interviews (the viability and sustainability of agriculture in rural communities is discussed to a certain degree);

3. factors other than provincial policy that influence severance activity, including proximity to urban centres, population density, livestock concentrations, costs of community services, and prices for land and agricultural commodities;

4. trends in non-residential development in agricultural designations; and

5. specific policy recommendations based upon interviews and study findings.

PLANNING AND LAND DIVISION IN ONTARIO

Since 1978, a succession of policies has protected agricultural land via limits on land division: namely, the Foodland Guidelines (1978), Growth and Settlement Policy Guidelines (1992), Comprehensive Provincial Policy Statement (PPS), three subsequent versions of the PPS (1994, 1996, 2005), the Places to Grow Act and Plans (2005), the Greenbelt Act and Plan (2005), and corroborating versions of the Planning Act (1990) (Sinker 2009).

All municipal plans must *be consistent with* the 2005 PPS, which maintains agricultural land protection as a provincial interest. The 2005 PPS defines prime agricultural areas as those in which soil classes 1, 2, and 3 predominate. Surplus dwellings are the only remaining type of residential lot creation permitted in prime agricultural areas (2005 PPS, Section 2.3.4.1). Prior to the 2005 PPS, municipalities were also permitted to allow retirement lots and infill lots, depending on policies implemented through the pertinent official plans.

Currently, Section 50 of the Planning Act (1990) permits municipalities to divide land in the form of consent for subdivision (many lots) or severance (typically no more than three lots). Part-lot control exemptions are also used to rearrange or create lots within pre-existing registered plans of subdivision. Upper- and single-tier municipalities in Ontario currently have consent-granting authority, unless this authority has been delegated to a lower tier. This study focuses on severances (also known as consents) and does not discuss subdivisions or part-lot control exemptions.

THE SEVERANCE PROCESS

Municipalities usually administer their authority to grant consents through a bylaw that creates a committee structure. These committee structures take various forms across Ontario:

- Committee of the Whole (all councillors make a decision at committee and forward it to council);

- Land Division Committee (can combine volunteer community members, councillors, and staff);

- Committee of Adjustment (authority for both consent-granting and zoning variances); and a

- streamlined process whereby straightforward applications are delegated to staff.

Severances are typically granted with a number of conditions, including rezoning to prohibit a residence on the remnant farm parcel, various surveys and inspections, and agreements to be registered on title.

Consent applications are first circulated to neighbours, related ministries, and community agencies. The designated consent-granting authority (committee or otherwise) then considers the application and public comments. The right to appeal to the Ontario Municipal Board (OMB) is granted to agencies and any party that has participated in the process. A diagram of the land division process is available from the Ontario Ministry of Municipal Affairs and Housing (OMMAH) website.

RESEARCH RESULTS
Results for the Province as a Whole
Ontario municipalities with agricultural designations received 113,695 severance applications outside major or separated urban centres from 1990 to

2009.[1] Of this number, 70,936 applications occurred from 1990 to 1999, and 42,759 occurred from 2000 to 2009. Compared between policy periods, there were 7,515 applications per year from 1990 to 1996, 4,923 applications per year from 1997 to 2005, and 3,662 applications per year from 2006 to 2009. This represents a 35 percent decrease in overall severance applications per year in Ontario following the 1996 PPS and another 26 percent decrease following the 2005 PPS.

Out of the total applications, 16,475 created new residential lots in agricultural designations. Of these applications, 11,552 occurred from 1990 to 1999, and 4,923 occurred from 2000 to 2009. Compared between policy periods, there were 1,309 residential lots created per year from 1990 to 1996, 687 lots per year from 1997 to 2005, and 284 lots per year from 2006 to 2009.[2] This represents a 48 percent decrease in new residential lots created in Ontario's designated agricultural areas when comparing before and after the 1996 PPS and an additional 59 percent decrease following the 2005 PPS compared with the 1997–2005 levels. The decrease in residential lots in agricultural designations was roughly double the decrease in overall severance applications in all designations, with 1.4 times the decrease after the 1996 PPS and 2.3 times the decrease after the 2005 PPS. This suggests that PPS policies had proportionally greater impacts on severances in agricultural designations than in other designations.

One of the most significant changes in 2005 was to no longer permit retirement lots or infill lots. The current study for 2000–09 shows a steep decrease in the number of retirement lots and infill lots created after 2005. In the four years of 2001–05, 1,045 retirement and infill lots were created; in the four years of 2006–09, only forty-five such lots were created. The results also show a decrease in farm help lots after 1996. Surplus dwelling severances continued at relatively similar rates before and after 2005.

Cumulative Severance Activity

Although the cumulative number of new lots in the agricultural designation continues to climb, the number of lots created per year is slowing. Figure 8.1 illustrates this trend from 1990 to 2009. The number of new lots created before 1990 is unknown, though numbers circulated to OMAFRA from 1979 to 1995 indicate much higher numbers of consent applications (Caldwell 1995). Although the true quantity of lots in Ontario's agricultural designation is much higher than illustrated here, because of lots created before the study began in

Figure 8.1. Cumulative new lots in the agricultural designation, 1990–2009.

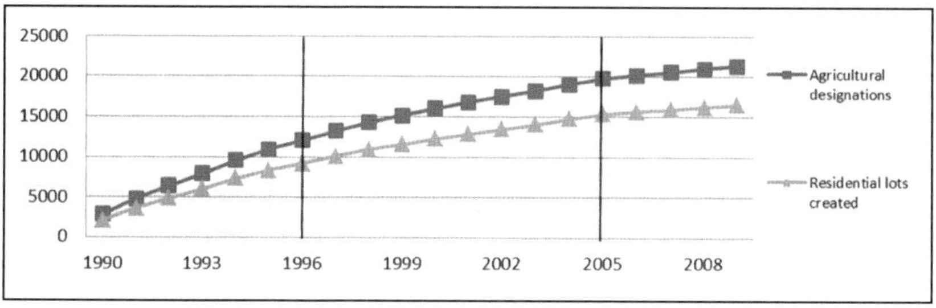

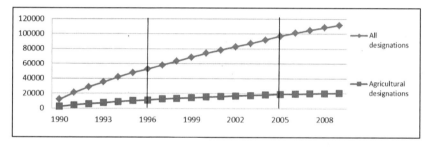

Figure 8.2. Cumulative total new lots in all designations, 1990–2009.

Figure 8.3. Retirement and surplus lots created per year, 1990–2009.

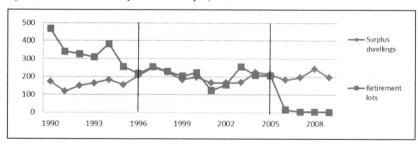

Figure 8.4. Farm help and infill lots created per year, 1990–2009.

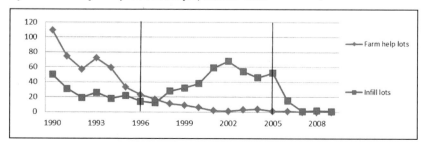

1990, the overall trend is clear: although the cumulative number of severances continues to grow, severances in the agricultural designation are decreasing.

The number of lots created in agricultural designations is increasing at a much slower rate than the overall number of new lots in all designations (see the growing gap in Figure 8.2). This indicates that less change is occurring in agricultural designations compared with non-farm rural and urban designations. This can be considered a desired impact of provincial policies, which have redirected new residential and non-farm developments to settlement areas, though other factors work alongside provincial policies to encourage this trend.

Rural Non-Farm Development by Type

The 2005 PPS specifically permitted farm splits and dwellings rendered surplus as a result of farm consolidation (PPS 2.3.4.1). It no longer listed infill and retirement lots as types of lots that municipalities were permitted to create. The effects of these policy changes are clearly visible in the following graphs. Figure 8.3 describes the decrease in the number of retirement lots created per year after the 2005 PPS changes. In the same graph, surplus dwellings continue without significant change after 2005.

In Figure 8.4, the decline of farm help and infill lots is also evident. The trend in farm help lots seems to have spiked in response to the Growth and Settlement Policy Guidelines introduced in 1992 and then declined quickly after the Comprehensive Provincial Policy Statement (CPPS) introduced in 1994. In comparison, infill lots were still permitted in the 1996 PPS (after being introduced in the 1994 CPPS) and continued with an upward trend per year until the 2005 PPS, after which such lots also quickly trended toward zero per year.

The results illustrated in the preceding graphs suggest that provincial policies have influenced severance trends in the ways intended.

Results for Each Provincial Region

Of the total new lots of all types (see Figure 8.6), 16 percent were created in central Ontario, 27 percent in eastern Ontario, 42 percent in southern Ontario, and 15 percent in western Ontario (see Figure 8.5 for regions).

Most notably, southern Ontario has a high proportion of residential development tied to farm uses. Western Ontario has the overall lowest residential lot creation and an overall lower rate of severances than other regions of the province. The highest quantities of non-farm residential lots occurred

Figure 8.5. Geographic regions of Ontario.

Figure 8.6. Total lot creation by type across Ontario regions, 1990–2009.

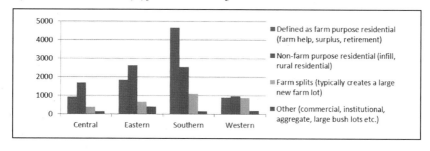

in southern and eastern Ontario. Eastern and central Ontario had higher proportions of non-farm residential development, often in the form of recreational and seasonal lots. To some degree, this illustrates an issue in determining which lots to include in the study, because cottage areas might have been designated agricultural, rather than lakeshore or recreational, in the same way that these areas have been designated in other parts of the province.

ANALYSIS OF RESEARCH RESULTS
Comparing Municipal Lot Creation Rates

When comparing lot creation trends among municipalities, it is important to account for differences in land area. Each municipality is responsible for an annual report of the number of acres designated for agriculture within its boundaries. This number is part of the provincial initiative to collect Municipal Performance Measures (MPM). The most recent MPM number for each municipality is used to represent the prime agricultural area available for lot creation. The most recent and consistently available MPM number for agriculturally designated land is from 2007 (MPM 2007).

The following formula, used originally by Caldwell (1995), is used to calculate ratios to compare residential lot creation trends in municipalities across the province:

Lot creation ratio = total number of residential lots created per period (including estimates)

(Total acres of land designated agricultural in 2007 MPM) / (1,000 acres)

The ratio describes the number of residential lots created in a typical concession block of 1,000 acres in a defined period of years. This number helps to illustrate the impact of residential development among municipalities with different land areas. The higher the number, the more restricted agriculture will be (most true for livestock production). For reference, all ratios across the province are provided in Tables 8.1 and 8.2.

On average, between 1990 and 1999, there were 1.64 residential lots created per 1,000 acres in the province. From 2000 to 2009, there was an average of 0.64 lots created per 1,000 acres of agriculturally designated land (this number can be referred to as the ratio). The shift from 1.64 to 0.64 in one decade represents an overall decrease of 61 percent or an average decrease of 53 percent across all municipalities. This average should be interpreted with caution because of wide variation across the province. The highest ratio was Prince Edward (7.55), whereas the lowest ratio was Perth (0.08). In a histogram of the data (see Figure 8.7), eight municipalities created less than one lot per 1,000 acres, twelve created between one and two lots per 1,000 acres, four created between two and three lots per 1,000 acres, and six created between three and four lots per 1,000 acres. This accounts for thirty of thirty-five municipalities. The remaining five municipalities created more than four lots per 1,000 acres from 1990 to 2009 (see Figures 8.8 and 8.9 for illustrations of the geographic distribution of these lots).

Changes to the PPS in 1996 and 2005 essentially created three provincial policy periods: 1990–96, 1997–2005, and 2006–09. Other significant policy changes could foreseeably influence severance trends. They are the introduction of Growth and Settlement Policy Guidelines of 1992; the Comprehensive Provincial Policy Statement of 1994; the amalgamation or redefined boundaries of a number of municipalities from 1997 to 2002; the delegation of consent-granting authority around the same time period from several upper tiers to lower tiers; and the introduction of a number of new official plans, often changing significantly within several years of the most recent PPS. Average lot creation per year in each policy period is a number that helps to explore some of the variation in lot creation over the twenty-year study period. On average, municipalities created thirteen fewer lots per year after the 2005 PPS.

Figure 8.7. Frequency of severance ratio values among municipalities.

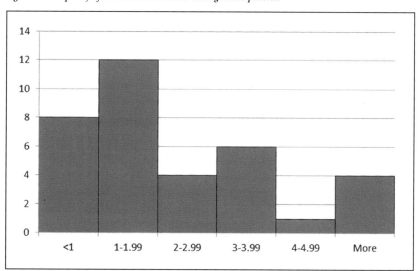

Table 8.1. Residential lots created per 1,000 acres compared between 1990–99 and 2000–09.

MUNICIPALITY	1990–99	2000–09	% CHANGE	OVERALL 1990–2009
PERTH	0.08	0.01	-89	0.08
WATERLOO	0.19	0.07	-65	0.25
MIDDLESEX	0.20	0.10	-49	0.31
OXFORD	0.22	0.10	-56	0.31
HURON	0.21	0.21	0	0.42
BRUCE	0.34	0.24	-31	0.58
DURHAM	0.42	0.37	-12	0.79
LAMBTON	0.65	0.30	-54	0.95
YORK	0.80	0.20	-74	1.00
ELGIN	0.48	0.58	+21	1.06
WELLINGTON	0.94	0.20	-78	1.14
SIMCOE	0.75	0.40	-46	1.15
HALDIMAND	0.81	0.37	-54	1.18
GREY	1.26	0.05	-96	1.31
BRANT	0.68	0.65	-4	1.34
HALTON REGION	1.15	0.30	-74	1.45
KAWARTHA LAKES	1.38	0.24	-83	1.61
DUFFERIN	1.37	0.53	-61	1.90
PETERBOROUGH	1.14	0.77	-32	1.91
CHATHAM-KENT	0.96	1.00	+4	1.96
RENFREW	1.56	0.46	-71	2.01
STORMONT-DUNDAS-GLENGARRY	1.55	0.58	-62	2.13
PRESCOTT-RUSSELL	1.97	0.51	-74	2.48
NORTHUMBERLAND	2.11	0.45	-79	2.56
NORFOLK	1.96	1.08	-45	3.04
HAMILTON	2.67	0.64	-76	3.31
PEEL (TOWN OF CALEDON)	2.93	0.43	-85	3.36
OTTAWA	2.24	1.30	-42	3.55
NIAGARA	2.47	1.13	-54	3.60
HASTINGS	2.74	0.87	-68	3.61
LENNOX-ADDINGTON	1.95	1.85	-5	3.81
LEEDS-GRENVILLE	4.53	0.65	-86	5.18
LANARK	5.66	0.40	-93	6.06
ESSEX	3.61	3.17	-12	6.78
PRINCE EDWARD	5.40	2.15	-60	7.55
AVERAGE ACROSS ONTARIO	1.64	0.64	-53	2.28

Source: Caldwell, Churchyard, Dodds-Weir, Eckert, and Procter (2011). *Lot Creation in Ontario's Agricultural Landscapes: Trends, Impacts, and Policy Implications.* Available at www.waynecaldwell.ca/projects/ruralnonfamrevisit.html.

Table 8.2. *Average number of lots created per year in each policy period.*

MUNICIPALITY	1990–96	1997–2005	2006–09	CHANGE IN AVERAGE NUMBER OF LOTS PER YEAR AFTER 2005*
PERTH	5	0	0	0
WATERLOO	2	4	1	-3
MIDDLESEX	NA	12	4	-8
OXFORD	8	9	0	-9
HURON	13	13	16	+3
BRUCE	20	13	14	+1
DURHAM	12	15	3	-12
LAMBTON	36	14	9	-5
YORK	8	4	1	-3
ELGIN	22	24	15	-9
WELLINGTON	45	15	5	-10
SIMCOE	NA	33	12	-21
HALDIMAND	26	13	7	-6
GREY	NA	1	2	+1
BRANT	NA	17	5	-12
HALTON REGION	7	1	1	0
KAWARTHA LAKES	55	15	4	-11
DUFFERIN	35	19	2	-17
PETERBOROUGH	20	11	6	-5
CHATHAM-KENT	54	64	49	-15
RENFREW	16	7	3	-4
STORMONT-DUNDAS-GLENGARRY	95	40	30	-10
PRESCOTT-RUSSELL	69	19	13	-6
NORTHUMBERLAND	50	11	1	-10
NORFOLK	55	32	18	-14
HAMILTON	41	18	3	-15
PEEL (TOWN OF CALEDON)	24	4	2	-2
OTTAWA	60	41	12	-29
NIAGARA	81	47	8	-39
HASTINGS	45	11	4	-7
LENNOX-ADDINGTON	23	11	0	-11
LEEDS-GRENVILLE	98	21	6	-15
LANARK	68	13	1	-12
ESSEX	120	124	33	-91
PRINCE EDWARD	56	27	10	-17
AVERAGE ACROSS ONTARIO	42	21	8	-13

Source: Caldwell, Churchyard, Dodds-Weir, Eckert, and Procter (2011). *Lot Creation in Ontario's Agricultural Landscapes: Trends, Impacts, and Policy Implications.* Available at www.waynecaldwell.ca/projects/ruralnonfamrevisit.html.

* The final column is the average number of lots created per year from 2006 to 2009, subtracted from the average number of lots created per year from 1996 to 2005; a large negative number signifies that a county created far fewer lots per year from 2006 to 2009 than it had from 1996 to 2005.

146 Farmland Preservation

Figure 8.8. Distribution of residential lots created per 1,000 acres, 1990-2009.

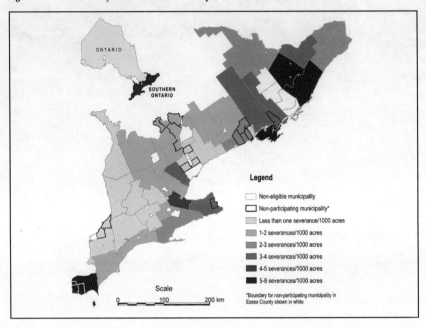

Figure 8.9. Percentage change from 1990-99 to 2000-09 in lots created per 1,000 acres.

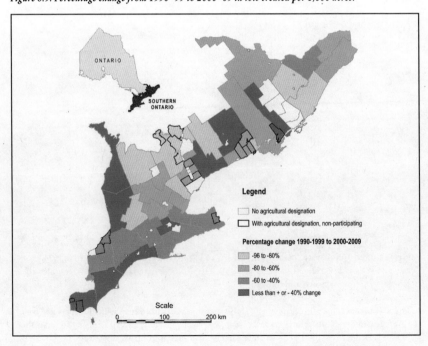

From the previous pages, the clearest trend is that the two decades differ markedly in the ratio of lots created per 1,000 acres. The decrease occurred in all municipalities, often dramatically, with the exception of Huron, which remained unchanged, and Chatham-Kent and Elgin, both of which increased. In both decades, lot creation ratios were the highest in a cluster of municipalities around Ottawa in eastern Ontario; around Prince Edward County; in the Greater Golden Horseshoe, including Hamilton and Niagara; and in the extreme south, Essex and Chatham-Kent.

Figure 8.9 clearly illustrates that the municipalities most impacted were rural metro-adjacent and intermediate municipalities. These areas are loosely arranged in horseshoes around Ottawa, Hamilton, and Toronto. Caution should be taken in analyzing this map, however. Large percentages resulted in some places that already have low numbers of severances. For example, even though Perth is marked as changing 89 percent, in reality very few severances were permitted before 2000, and virtually none was permitted after that year, which creates a large percentage but a small change in absolute terms. The change in absolute terms can be seen in Table 8.1.

Individual County and Regional Observations

The 1996 and 2005 PPS changes had significant impacts on lot creation across Ontario. On average, each municipality created forty-two lots per year from 1990 to 1996, twenty-one lots per year from 1997 to 2005, and eight lots per year from 2006 to 2009. These numbers provide a basis for understanding how policies have effectively reduced residential lot creation while allowing cumulative impacts from residential lots to increase. Accounting for the differences in land size among municipalities, on average 2.28 lots were created per 1,000 acres of agricultural designation in municipalities across the province from 1990 to 2009.

Results vary widely among municipalities. Very little change occurred for municipalities in western Ontario, which tended to have low severance activity throughout the study period. In southern Ontario, many municipalities drastically reduced residential lot creation. For example, Essex County decreased residential lot creation by ninety-one lots per year from 2006 to 2009. Niagara Region decreased residential lot creation by thirty-nine lots per year from 2006 to 2009. Eastern Ontario tended to create a high number of lots relative to its smaller areas designated agricultural. Eastern and central Ontario had more lots created for non-farm purposes than other parts of the province.

Perth County had the lowest number of lots created per 1,000 acres in the province (0.08); the county is composed of predominantly prime agricultural land, with a high concentration of livestock. Neighbouring Waterloo Region was the second lowest (0.25), notable given the combination of productive agricultural areas adjoining areas facing high pressure for urbanization. Prince Edward County had the highest number of lots created per 1,000 acres (7.55). One possible explanation is the proximity of many agricultural designations to the lakeshore in the county. Essex County, located at the most southwestern corner of the province, had the second-highest number of lots created per 1,000 acres (6.78), despite a large reduction in recent years.

Twelve of thirty-five upper-/single-tier municipalities now permit fewer than two severances per year in the agricultural designation. In these municipalities, the combination of local and provincial policies has created a strong policy precedent in which very few new lots can be created. Conversely, three municipalities continue to permit more than thirty new lots per year (Chatham-Kent, Essex, and Stormont-Dundas-Glengarry).

These are general observations, and there are exceptions to the rule. Full results are available for each municipality and each of Ontario's geographic regions in "Report 2: Profiles and Summaries" (www.waynecaldwell.ca).

Assessment of the Impact of Rural Non-Farm Development on Livestock Agriculture

Residential lot creation in agricultural designations produces numerous impacts. These impacts occur in a variety of forms, including net costs for municipal servicing, loss of agricultural land, restrictions on a range of agricultural commodities, shifts in political and social values, conflicts with neighbours, and lost opportunities in the future value of farmlands (Caldwell and Weir 2002). Although space does not permit the exploration of each impact here, one informative illustration can be made in the case of livestock agriculture.

In total, 16,475 new residential lots were created in agricultural designations from 1990 to 2009. For the purpose of illustration only (not as a definitive measure of impact), consider that each additional lot will create a circle around it in which no livestock expansion is permitted as per minimum distance separation requirements under the province's Nutrient Management Act (2002) and Provincial Policy Statement (2005). As a hypothetical case, consider a residential lot being severed in a prime agricultural area. About 40 percent of new or expanded barns were between 100 and

500 nutrient units in the period 2001–06, as measured by building permits in the most livestock-intensive part of Ontario (Eveland et al. 2005). Based upon the current minimum distance separation for Type A (low-density) residential uses, a distance of several hundred metres would be required for a barn in the range of 100–500 livestock units, depending on livestock type and manure storage method. For illustrative purposes, consider a minimum distance separation of 250 metres. The area of land in the minimum distance separation circle around the property would have a minimum area of 48.5 acres. Multiplied by the number of residential severances from 1990 to 2009 (16,475), that would result in 799,349 acres of land being incapacitated for much of new livestock agriculture investment. Although this is by no means an accurate estimate of the impact, it is helpful in illustrating the scale of cumulative impact that might well be a result of historical and ongoing severance activity.

CONCLUSION

The results of this research demonstrate that current restrictions on new residential lot creation are closing the gap between provincial policies outlined in the PPS and the reality of ongoing residential lot creation in the countryside. This finding should be informative for policy makers seeking evidence of the effectiveness of broad restrictions on land division through planning policies.

There continues to be a large number of residential severances in the countryside, though at a slower rate than at the beginning of the study period. Cumulative lot creation data will be helpful in further research that would outline potential economic and geographic factors as well as estimate the impacts of ongoing severance activity. Using these data, policy makers can assess the overall impacts rather than consider severance applications on a case-by-case basis only. When a single severance is considered on its own merits, it might seem as though private benefits outweigh public costs of residential development. The argument for farmland preservation is bolstered through fact-based cumulative assessment of number of lots and associated impacts.

When cumulative factors and impacts are understood, policy makers can begin to support further landscape-scale planning policies and decision-making structures. The provincially led planning approach to identifying and protecting prime agricultural areas at the landscape scale is an important progression toward addressing the gap between private property interests and public values.

This research strives to illustrate key issues within planning policies for lot creation in agricultural designations of Ontario. Based upon the effectiveness of new lot restrictions introduced in the 2005 PPS, it is apparent that continued planning reform is contributing to more sustainable use of prime agricultural lands. At the same time, greater communication of the policy rationale is required from the province to municipalities, and from municipalities to landowners, if policy reforms are to be implemented to the greatest effect possible in local decision making.

NOTES

1   With thanks to the research assistance of Anneleis Eckert, Charlie Toman, and Kate Procter.

2   Note that aggregate statistics at the provincial level contain estimates for years in which insufficient data were available for specific counties. All data for the years 1990–95 were estimated for Middlesex, Grey, Brant, Haldimand, and Norfolk. These estimates were based upon multiplying the total severance application numbers retrieved for each county from OMAFRA (1995) by the average percentages of each lot creation type in each county. Additionally, data for some in-between years were estimated in Hastings, Peterborough, and Stormont-Dundas-Glengarry (for several years from 2000 to 2009) and Kawartha Lakes, Prince Edward, Lennox Addington, and Hamilton (for several years from 1990 to 1999). In-between years were estimated by multiplying the total application numbers obtained from the municipality by the average percentage of each lot creation type in the years directly before and after the estimated year. Simcoe was not included in aggregate statistics at the provincial level because of lack of data.

REFERENCES

Bryant, C.R., and L.H. Russwurm. 1979. "The Impact of Non-Farm Development on Agriculture: A Synthesis." *Plan Canada* 19 (2): 122–39.

Caldwell, W.J. 1995. "Rural Planning and Agricultural Land Preservation: The Experience of Huron County, Ontario." *Great Lakes Geographer* 2 (2): 21–34.

Caldwell, W., and C. Weir. 2002. "Ontario's Countryside: A Resource to Preserve or an Urban Area in Waiting?" School of Environmental Design and Rural Development, University of Guelph. http://www.waynecaldwell.ca.

Eveland, C., A. Weersink, W. Caldwell, and W. Yang. 2005. "Spatial Trends in Barn Building Permits." *Great Lakes Geographer* 12 (1): 19–27. http://geography.uwo.ca/research/the_great_lakes_geographer/docs/Volume%2012/2_EvelandEtAl.pdf.

MPM (Municipal Performance Measures). 2007. "Effectiveness Categories: Number of Hectares Redesignated from an Agricultural Use, and Building Lots Permitted outside a Settlement Area." http://csconramp.mah.gov.on.ca/fir/Welcome.htm.

Red Deer County. 2006. "Cost of Community Services Study for Red Deer County." http://www.rockies.ca/programs/cocs.htm.

Sinker, G.E. 2009. "Agricultural Consents and Land Use Planning." Ontario Bar Association Continuous Learning Series.

Statistics Canada. 2013. "Ontario Farm and Farm Operator Data, 2011." http://www.statcan.gc.ca/pub/95-640-x/2012002/prov/35-eng.htm.

CHAPTER NINE

Preserving and Promoting Agricultural Activities in the Peri-Urban Space

NICOLAS BRUNET

Many describe farming in the peri-urban countryside as farming in a land of uncertainty (Brunet 2006; Bunce and Maurer 2005; Esseks et al. 2010). It is a dynamic space where change is the norm and agricultural activities compete with pressures for urbanization, turning once intact rural communities into areas of transition and instability. This chapter explores the potential of preserving agricultural activities in the rural-urban interface through comprehensive strategies involving the state and, most importantly, the farmer.

A review of empirical studies in Canada and abroad reveals the potential of better farming models for preserving and promoting agriculture around our major urban centres. The chapter presents literature discussing the key characteristics of these models and the strategies that can lead to a thriving peri-urban agricultural industry. However, in attempting to understand how and why agriculture must change to survive, I first explore the major limitations to this important livelihood and economic activity along with the varying definitions of the term *peri-urban*.

DEFINING PERI-URBAN

Areas surrounding urban centres have been defined and named in various ways over time, including "peri-urban," "urban shadow," "urban field," "urban fringe," and so on. However, some authors make no attempt to define this ambiguous term and use it as though its definition is implicit (Sharp and Smith 2004). Iaquinta and Drescher (2000) conducted an extensive review to understand the commonalities among the various interpretations of this notion. Some examples follow:

- Peri-urban is different from urban.
- Peri-urban is, in some fashion, connected to being urban.
- Peri-urban has geographic (near the city), demographic (increasing population), and temporal (constantly expanding, changing) components.

In general, there is no single definition of this concept. In Fitzsimons (1983, 298), the *urban fringe* is defined as "a zone up to ten miles wide surrounding the suburbs where rural land is being turned into housing and industrial subdivisions." The *urban shadow* is an area "extending at least twenty to thirty miles beyond the urban fringe"; "it is still rural but changes are beginning to appear." This definition is very limiting since it defines the exact size of the area. Conversely, Daniels (1999, 9) defines the *fringe* as "a hybrid region no longer remote and yet with a lower density of population and development than a city or suburb." Another characteristic of the fringe has been the existence of relatively rapid population growth over the past several decades (Heimlich and Anderson 2001). Evidently, the rural-urban interface is a complex setting where study of the impacts of growth and change has been given a significant amount of research attention (AFT 1997). A more recent study attempts to define peri-urban more systematically by using specific ecological indicators, which can be assessed using geographic information systems and extensive field surveys (Macgregor-Fors 2010).

FARMING ON THE FRINGE: WAR ZONE OR LAND OF OPPORTUNITY?

The concern for farmland preservation has long been suppressed by the common thought that land in Canada is limitless and that, if needed, technology would make it possible to grow large amounts of food on relatively small areas of land (Bunce 1985). Only recently have provincial and federal governments shown an interest in the maintenance of agricultural land uses under threat of urban expansion. Now most jurisdictions have some form of farmland preservation policy in place (Beesley 2010; Furuseth and Pierce 1982). This movement has also been influenced by the recognition that some of the best agricultural lands in areas of urban expansion are also the most attractive for non-agricultural uses (Bunce 1985). This attraction combined with growing suburban populations has further increased the demand for development at the city's edge, leading to speculation and the purchase of large tracts of farmland by mainly absentee non-farmers who rent out the land until development ensues (Bunce 1985). In a study by Brown, Phillips, and Robert (1981), a survey was conducted around two major Canadian cities, and it was determined that 80 percent of the land was owned by investors and developers. This has both negative and positive impacts on agriculture in the area. First, as land prices rise in expectation of urban development, it is known that investment in agriculture will decline (Sinclair 1967). Furthermore, lowering the percentage of farmer-owned land because of high prices and lack of availability causes fragmentation and insecurity. Consequently, stability of the agricultural industry becomes uncertain (Bunce 1985).

Ownership of the land ensures not only this stability but also the sustainability of the industry by influencing use of the land by farmers and lowering incidences of intensive cash cropping (Kormacher 2000). However, though the price of land is very high, farmers recognize that land rental is usually inexpensive and allows them to lower their capital costs, making it possible for them to expand when needed (Mann 2002). This is further developed in Bryant and Fielding (1980), in which farmland rental from non-farm landowners was found to be a significant factor in agricultural development of the urban fringe. In a more recent study of farming in the Greater Toronto Area (GTA), Bunce and Maurer (2005) found that the land market, the possibility of retaining development rights, and the possibility of selling land to the highest bidder were very important to farmers. High land prices were believed to be a strength more than a weakness. They do increase land assessments and taxes, however, which some believe are becoming unaffordable (Mann 2002).

Expanding urban areas also bring another threat to the surrounding countryside: *people*. In an article by Mann (2002), the influences of suburban development on surrounding farming activities in the GTA are explored. A major impediment is traffic. Some farmers are unable to reach now distant rental lands at peak traffic hours or find it very difficult to move large machines on main roads. GTA farmers are also faced with complaints from non-farm neighbours about smell and noise. In some regions, spraying is limited to certain hours, further limiting the agricultural operation. The incidences of crop theft, garbage dumping, and trespassing have also increased, making it difficult to want to maintain a farming operation in these areas. It has also been expressed that constant proposals for new municipal bylaws aimed at pleasing suburban residents affect farmers (Brunet 2006; Mann 2002). Loss of the agricultural community and rural character in the peri-urban space further limits the future of farming. According to Bunce and Maurer (2005), most farmers believe that they are not adequately represented at the municipal level, and many think that the local farming community no longer exists or is in serious decline. Overregulation is also perceived by many as a major threat to their continuing farming. Most farmers are also concerned about having to adjust to operating next to non-farm neighbours. As in the article by Mann (2002, 37), some informants have received complaints about "smell, noise, and pollution from livestock operations, fertilizer and manure spreading, pesticide spraying, dust from field preparation and cultivation" (also see Bunce and Maurer 2005).

Most importantly, decline in the number of farms because of instability in the near-urban agricultural industry might eventually cause specialized feed, fertilizer, and seed suppliers to leave the area (Katz 1997). A reduction in vendors might also result in higher prices charged by the equipment dealers and repair services that remain. This loss of farm support structures also fuels uncertainty. Therefore, farmers might hesitate to invest in maintaining infrastructure, resulting in further decline of the local agricultural industry and community.

HOW DO WE SAVE PERI-URBAN AGRICULTURE?

To preserve both farmland and agricultural activities, many strategies have been proposed and utilized throughout North America (Beesley 2010; Esseks et al. 2009). For the purpose of this discussion, these strategies are divided into two main categories: regulatory/incentive-based approaches and farm-level approaches. Whereas regulatory and incentive-based techniques are generally implemented by the government or other agencies, farm-level approaches to farmland preservation are initiated by farmers or local farm organizations. My focus here is on the role of operators in adapting their agricultural systems in response to the many forces of change acting at the farm level. These forces include not only threats but also opportunities that have been found to promote innovation and therefore create distinct farm models in the peri-urban space (Heimlich and Anderson 2001; Lapping and Pfeffer 1997; Zasada 2011) and support the adaptive capacity of food producers (Bryant et al. 2009).

Regulatory/Incentive-Based Approaches

The development of farmland protection programs in Canada began in the early 1970s. At the time, concerns were raised about the uncontrolled expansion of cities on prime farmland. British Columbia (Agricultural Land Reserve, 1973) and Quebec (la Loi sur la Protection du Territoire Agricole, 1978) were the first provinces to adopt farmland protection programs. These programs were subject to much public debate and controversy. It was found that such large-scale protection programs, though very important, did not have the predicted positive impacts on farmland preservation. Focused mainly on preserving *the land*, these efforts simply overlooked regional issues, including farm viability and needs of the farmer to maintain a livelihood. Adoption of the Greenbelt Act and Greenbelt Plan (2005) and Growth Plan for the Greater Golden Horseshoe (2005) in central Ontario were yet other attempts by a provincial government to save agricultural land from development. The full impacts of this initiative are not yet understood, and farmers are concerned that the plan does not include the promotion of farm viability (OFA 2005). Other concerns include the lack of published criteria to address inefficient land use and speculation as well as shrinking opportunities for publicly accessible land, lack of an effective appeal process for local zoning decisions and a clear scientific basis for delineation of the outer greenbelt boundary, as well as no mechanism to coordinate land use across upper-tier municipal boundaries within this greenbelt (Fitzsimons et al. 2012). Clearly, the protection and maintenance of viable agricultural activities transform a land-use planning issue into a societal issue, involving stakeholders at both the provincial and the local levels.

These policy-driven initiatives demonstrate that the regulation of land use is evidently the most common tool for the conservation of agricultural land in Canada. However, these programs are generally a combination of many techniques that form a comprehensive whole. They include regulatory/incentive-based approaches (AFT 1997).

Regulatory approaches include agricultural protection zoning, cluster zoning, and mitigation ordinances and policies. Zoning is an extensively used land-use control (Fitzsimons et al. 2012; Peterson 1982). However, this method is also relatively easy to change and often subject to political pressures related to development. This technique offers no assurance that an agriculture-zoned area will remain so in the near future. Mitigation ordinances and policies are a relatively new farmland protection technique and not widely used. They require developers to permanently protect one acre of farmland for every acre of agricultural land that they convert to other uses (AFT 1997). This is generally achieved with the use of agricultural conservation easements on farmland in other parts of the city.

In *Saving American Farmland: What Works* (AFT 1997), various incentive-based techniques for farmland preservation are also explored. They include differential tax assessment laws, circuit breaker tax relief credits (United States), right to farm laws, conservation easements and purchase of conservation easement programs, transfer of development rights (United States), and agricultural districts. Many of these initiatives are currently being developed or have been established in Canada. They are explored further below.

Differential tax assessment laws are widespread throughout Canada. They help to promote the viability of agriculture by reducing the amount of money that farmers are required to pay in local property taxes (AFT 1997). Historically, in the United States and Canada, farmland was assessed at its "highest and best use," usually meaning for urban purposes. Urban fringe farms were then forced out of business because of increasingly higher taxes (Peterson 1982). Today in Ontario most farmers pay approximately 25 percent of residential rates on their farmlands. If not, it is agreed that farmers would pay a disproportionately high share of taxes compared with their demand for local public services (AFT 1997). Therefore, differential tax assessment laws direct local governments to assess agricultural land at its value for agriculture and not at its value in the speculative market.

Right-to-farm laws are found throughout Canada and the United States (Esseks et al. 2009). They are intended to protect farmers from nuisance lawsuits. In some instances, statutes protect farmers from lawsuits filed by neighbours who arrived after the agricultural operation was established. Otherwise, farmers are protected as long as they use accepted management practices in compliance with the law. Right-to-farm legislation also prohibits the implementation of local ordinances that would impose restrictions on normal agricultural activities.

Agricultural conservation easements and the purchase of agricultural conservation easement (PACE) programs are used extensively in the United States. Conservation easements are voluntary legal agreements between a private landowner and a land trust, conservation organization, or government agency. Landowners can receive tax benefits as a result of their donations. The trust or other organization is therefore responsible for enforcement of the easement and monitoring use of the land (AFT 1997). Conversely, PACE programs are designed to pay farmers to protect their lands from development. In general, landowners sell easements to a government agency or private conservation agency. The farmer is usually paid the difference between the agricultural value of the land and the value of the land for its highest and best use (AFT 1997). The absence of land rights in the Canadian Constitution has limited the development of these costly programs.

The creation of agricultural districts is also gaining ground in the United States. By the mid-1990s, sixteen states had enacted agricultural district laws, up from seven in the 1980s (AFT 1997). Since then, six more states have enacted right-to-farm legislation at the regional level, closely emulating agricultural districts laws (AFT 2008). Their creation enables farmers to form special areas where agriculture is encouraged and protected (AFT 1997). Enrolment is voluntary, and farmers receive a range of benefits depending on the area. These benefits include automatic eligibility for differential assessments, protection from development pressures, and certain tax exemptions. Moreover, farmers are not assessed for utilities as long as they do not use them (Peterson 1982). However, agricultural districts do not ensure that lands under the most threat of urban expansion will be included in the district. These are voluntary programs and therefore tend to be ineffective when faced with market pressures (Peterson 1982).

The final large-scale farmland protection program, now across much of the United States and starting in Canada, is payments for environmental benefits (also known as ecological goods and services) models. These programs aim to compensate farmers for providing a wide range of services, from wooded areas to buffer strips along rivers. In the United States, there are countless examples of these programs. For example, the Watershed Agricultural Council (WAC 2005) is a non-profit organization funded by the City of New York, the USDA Forest Service, and other federal sources. Its mission is to support the economic viability of agricultural and forest industries through protection of water quality and promotion of the New York City Watershed region (WAC 2005). It provides farm-planning advice, conservation easement programs, and compensation for source water protection to farmers. In Canada, the proposed Alternative Land Use Services (ALUS) is an interesting framework. It was designed by the farm community to address the growing environmental concerns of Canadians. It aims to promote better stewardship of the land by farmers with compensation for environmental services provided to society. It is a voluntary program, and participants may be able to enroll up to 20 percent of their lands for a flexible time period specified by contract. Furthermore, ALUS intends to monitor and manage implementation of the program to ensure transparency. These programs will promote viability of the agricultural industry but not ensure the protection of agricultural land in areas where urban development pressures are strong. Would these payments be sufficient to offset the high costs of near-urban farming?

Lynch and Carpenter (2003) found that very few, if any, of the farmland preservation programs decreased the rate of farmland loss at the fringes of metropolitan areas. Several other studies have also suggested this. For the near-urban agricultural industry to survive, other approaches will therefore be needed. Some have suggested the integration of regulatory and incentive-based techniques (AFT 1997). Conversely, some studies have shown that the viability of near-urban agriculture will be ensured with the adaptation of production and marketing methods to meet the requirements of urban populations (Ilbery 1991; Sharp and Smith 2004). This calls for the birth of a new and innovative near-urban agricultural industry.

In a more recent study, Esseks et al. (2009) found that large minimum lot sizes discouraged property buyers not interested in farming, an urban growth boundary limited the expansion of city services into agricultural areas, and attractive rural scenery helped to convince many farmer and non-farmer residents to be politically active to protect their lifestyles rather than make money from selling their lands and moving elsewhere. In some cases, landowners voluntarily agreed to easements that limited non-agricultural uses on their lands, usually in perpetuity, in exchange for monetary compensation for the development rights thus surrendered.

Farm-Level Approaches

In response to the many challenges of farming at the rural-urban interface, some farmers have decided to adapt their production and marketing methods to ensure the viability of their operations (Bryant and Johnston 1992; Esseks et al. 2009; Heimlich and Anderson 2001; PlanScape 2003; Zasada 2011). The evolution of near-urban agriculture has been characterized by diverging thoughts and practices, which have led, on the one hand, to the maintenance of traditional practices and, on the other, to alternative agricultural models, often perceived to be better adapted to this environment.

Historically, near-urban farmers did not have access to efficient systems of transportation and generally specialized in fresh produce and livestock production to overcome problems related to conservation (Fleury and

Donadieu 1997). In the nineteenth century, Von Thunen proposed a model in which agriculture is organized around cities according to the ease or difficulty of transporting particular commodities to urban markets (Sinclair 1967). Farmers producing perishable commodities and other items difficult to transport were therefore located near the city's edge. At the time, cities had a vibrant fringe agriculture that supplied the urban market with most of its food and acted as a filter for urban pollution (Fleury and Donadieu 1997). Farming was then intimately linked with the urban environment. However, with the increase in transport efficiencies and heavy reliance on imports and the world market, this relationship, the basis of survival for both farmers and city dwellers, has now virtually vanished.

Over time, farmers have identified both limitations and opportunities in farming at the city's edge. Challenges are numerous and often lead farmers to move to where the pressures are less intense. However, despite these threats, there continues to be a substantial amount and diversity of farming activity in near-urban areas (Sharp and Smith 2004). It has been found that fringe farms are different from rural farms and show substantial variations in farm enterprise (Lapping and Pfeffer 1997). This growing shift in metropolitan agriculture is readily seen by planners and the public in land-use patterns (Scarfo 1990). In Bryant (1989), farm restructuring at the urban fringe came either from macro-changes to which individuals react and adjust or from individuals with more entrepreneurial and innovative behaviour.

FORCES OF CHANGE

In attempting to understand farm-level adaptation, it is first important to explore the major forces that promote this change. Although farmers have reacted to the limitations presented in an earlier section, the following reviews empirical studies on farm-level decision making and the underlying reasons for change.

First, Bowler, Bryant, and Nellis (1991) found that the geographic boundaries of the system of exchange, along with the characteristics of the exchange itself, changed according to three sets of forces. They describe these forces as the development of new needs, changing communication, and production technology.

New needs include changes in both consumer and producer behaviours. On the one hand, consumers are demanding *healthy* agricultural products, such as organic produce, initiating diversification in farming enterprises (Winter 2003). On the other, farmers are placing increasingly different values on different kinds of work, having impacts on production methods (Bowler et al. 1991). Moreover, new communication technologies have led to a global transformation in agricultural production. This has prompted the restructuring of farm management and marketing systems (Bowler et al. 1991). Furthermore, technological change in methods of production has historically played a major role in agricultural adaptation. Integration of the industrial model has led to the mechanization of production methods, capital substitution for labour, increasing reliance on non-farm inputs, and farm business consolidation.

Personal traits of the farmer (e.g., age and education) and household and farm income also play major roles in the restructuring of an agricultural operation (Bowler et al. 1991; Smithers and Johnson 2004). However, underlying processes in the adoption of an innovative strategy are difficult to understand and have led to various interpretations. One perspective assumes that farmers demand new innovations and that patterns of uptake will relate to the relative innovativeness of the adopter, which is heavily influenced by personal characteristics (Hagerstrand 1967). Another perspective claims that the ability to adopt an innovation is uneven. In this case, promotional communication and market segmentation might target only certain groups of farmers and farm types. Therefore, non-adoption is not the result of a lack of innovativeness but an issue of access to resources.

Furthermore, some believe that the decision maker actually looks for an optimal compromise among several objectives or attempts to achieve a certain satisfactory goal (Romero and Rehman 2003). For instance, a subsistence farmer might be interested in securing adequate food supplies for his or her family, maximizing cash income, and increasing the amount of time spent meeting social obligations. Likewise, a farmer in the "developed" world might typically be concerned with maximizing gross margins, minimizing debts, and increasing net worth (Romero and Rehman 2003). Therefore, it is clear that, independent of technological advances and consumer demands, many factors must be considered when analyzing the evolution of farming systems.

In Table 9.1, the forces acting on the peri-urban farm are divided into two major categories: those external and those internal to the operation and household. External forces are further subdivided into two categories: urbanization forces (specific to the near-urban space) and non-urbanization forces (not limited geographically). Urbanization forces include things such as high land prices and nuisance complaints, whereas non-urbanization forces include technological advances, climate, et cetera. They include both positive and negative impacts acting at all scales.

CHARACTERISTICS OF NEAR-URBAN AGRICULTURE

Clearly, the dynamic forces of urbanization create an interface where a variety of farm types coexist (see Table 9.1). In general, however, metropolitan farms have been "smaller, produce more per acre, have more diverse enterprises, and are more focused on high value production than non-metro farms" (Hoppe and Korb 2000, cited in Heimlich and Anderson 2001, 40; also see Bryant, Russwurm, and Wong 1984). Scarfo (1990) adds that metropolitan farms also tend to decrease capital and purchased input per dollar of output, shift from field and livestock crops to specialty or niche crops, increase the amount of contract and direct marketing, and be run by younger farmers who work more hours off the farm. In a Canadian study, Bryant et al. (1984) showed that near-urban areas contained high proportions of the most intensive agricultural activities, especially specialized crops, nurseries, and horticulture. Furthermore, Heimlich and Anderson (2001), in a study conducted by the USDA, define three distinct groups of "metro farms": *recreational*, *adaptive*, and *traditional*. In the same study, recreational operations or hobby farms accounted for 51 to 54 percent of total metro farms but were considered to have little economic viability. Sales were generally very low and had little ability to generate income for the farm household. Traditional operations accounted for approximately one-third of the total metro farms, whereas adaptive farms accounted for 13 to 14 percent. This last farm type was found to control "more than proportional shares of metro farm sales, assets and net cash farm income" (Heimlich and Anderson 2001, 41). The study also found that adaptive agricultural operations have the highest chance of survival in near-urban areas. Specifically, this group includes "farms that produce relatively high value products, with sales of $10,000[USD] or more and having sales of more than $500 per acre of land. Specializing in high value products allows these farms to adjust to increasing land prices, population density, and continuing conversion of local agricultural land to non-farm uses" (43).

Adaptive farmers have also been found to engage more readily in proactive behaviours, taking advantage of nearby expanding urban populations. In fact, opportunities in the peri-urban space have been explored extensively in recent years. Beauchesne and Bryant (1998, 321) found that "a certain degree of stress may have a beneficial impact in stimulating creative adaptation." They also found that urban pressures can promote innovation

Table 9.1. *A framework for peri-urban agricultural adaptation.*

EXTERNAL FORCES AT ALL SCALES		INTERNAL FARM FORCES	
URBANIZATION FORCES	**NON-URBANIZATION FORCES**	**BUSINESS**	**HOUSEHOLD**
• HIGH FARM EQUITY • NICHE MARKETS AND TOURISM POTENTIAL • PROXIMITY TO CONSUMER • OFF-FARM EMPLOYMENT • CULTURAL TRANSITION • CONSUMER TASTES • HIGH LAND PRICES • NUISANCE COMPLAINTS • RESTRICTIVE BYLAWS AND REGULATIONS • LACK OF POLITICAL SUPPORT AND REPRESENTATION	• CLIMATE/LATITUDE • COMMUNITY ORGANIZATIONS AND FARMERS' ASSOCIATIONS • NEIGHBOURS/ LOCAL CULTURE AND TRADITION • MEDIA/ACCESS TO INFORMATION • GLOBALIZATION • LACK OF QUALIFIED LABOUR • MARKET FLUCTUATIONS • PROVINCIAL REGULATIONS • BIOTECHNOLOGY AND OTHER TECHNOLOGICAL ADVANCES	• SOIL AND LAND QUALITY, LAND TENURE • OWNERSHIP, LOCATION • AGE OF OPERATION (ESTABLISHED VERSUS NEW) • DEBT LOAD • OPERATION TYPE AND CYCLE	• AGE OF OPERATOR • FAMILY CYCLE (AGE AND NUMBER OF CHILDREN) • INVOLVEMENT OF SPOUSE • EDUCATION • SUCCESSION PLANS • PHILOSOPHY AND FARMING BACKGROUND
METROPOLITAN FARM TYPES			
TRADITIONAL	ADAPTIVE		RECREATIONAL
FARM BUSINESS TRAJECTORIES			
INNOVATORS	PERSISTERS	LEAVERS	HOBBY FARMERS

Sources: This framework is inspired by Heimlich and Anderson (2001) and Smithers and Johnson (2004) and expands on Bryant and Johnston (1992).

throughout North America. Other positive impacts of urbanization on near-urban agriculture include better access to urban markets, proximity to specialized services and suppliers, and potential for direct sales to consumers. Heimlich and Anderson (2001) saw proximity to an abundant labour pool in an urban centre as another benefit.

ADAPTIVE STRATEGIES AND FARM BUSINESS TRAJECTORIES

Evidently, near-urban farming does have benefits, and in some cases farmers have acted on them. Ilbery (1991) found that the restructuring of fringe farm businesses is necessary. Farm adjustment strategies include changes in farm enterprise, labour, business structure, tenure, size, and diversification (Munton 1990).

Farm diversification generally refers to the development of alternative enterprises (Slee 1987). It implies the adoption of income-earning activities "outside the range of conventional crop and livestock enterprises associated with agriculture" (McInerney, Turner, and Hollingham 1989, 6). Therefore, it also involves a diversion of resources (land, labour, capital) previously committed to conventional farming activities (Bowler et al. 1991).

It is possible to distinguish two forms of diversification: *agricultural* and *structural* (Ilbery 1991; Zasada 2011). Structural diversification includes tourism, recreation, value-added production, alternative marketing, and methods of passive diversification, such as leasing lands or buildings. Agricultural diversification includes growing unconventional crops and raising atypical livestock, organic farming, energy farming, and agricultural contracting (Ilbery 1991). In some cases, farm diversification models include off-farm income generated from family members (Shucksmith et al. 1989). This broadened definition has been renamed *pluriactivity* by some (e.g., Gasson 1987).

Although the agricultural community has been known to discourage certain types of alternative development, the diversification of near-urban farming is growing throughout Canada. Many adaptive farms attempt to utilize the aforementioned strategies. However, agricultural adaptation in the near-urban context expands upon the traditional concepts of diversification and restructuring. In fact, in Sharp and Smith (2004), the integration of rural and urban interests was determined to play a key role in the survival of farming activities in this space. Described as an agroecosystem, the urban fringe is reliant for continued health on the establishment of avenues of communication between farmers and non-farm neighbours. Farmers who engage in "good neighbour" activities, including adapting fieldwork schedules to the needs of the surrounding population, have been able to reduce conflict and therefore enhance the sustainability of their operations (Chase and Hutcheson 1998). Communication can also be ensured with alternative marketing strategies. They include events, farm markets, pick-your-own days, and other methods aimed at increasing contact between consumers and producers (Ilbery 1991).

Undertaking unconventional enterprises has also been effective as an adaptive strategy (Esseks et al. 2009). In metropolitan areas, alternative crops and livestock are often high in demand but not faced with the pressures of the global market (Roberts 2005). Moreover, the increase in cultural and ethnic diversity in Canadian suburban areas has allowed some farmers to supply rare vegetables at premium prices directly to restaurants. For example, as mentioned in an article featured in *Better Farming* (Stoneman 2005), former cash crop farmers decided to switch to artichoke and other specialty crops, such as celery root. They are now the first commercial artichoke farmers in Ontario and have been very successful. Bunce and Maurer (2005) found that certain sectors thrive on being located close to residential areas. They include sod farms, pick-your-own operations, market gardens that undertake direct marketing, greenhouse operations, and so on. According to this study, "those who have diversified or moved into niche markets appear to have the most promising future" (40). Some consumers are also demanding local produce and are willing to pay for it. In response to this growing trend, community-supported agriculture (CSA) has emerged as an adaptive response to opportunities in the urban environment. The CSA model has been described as "an innovative and resourceful strategy to connect growers with local consumers, develop regional food supply and strong local economy, maintain a sense of community, encourage land stewardship, and honour the knowledge and experience of growers and producers working with small to medium farms" (Worden 2004, 322).

In Table 9.1, adaptive farmers are divided into four main categories: *innovators*, *persisters*, *leavers*, and *hobby farmers*. These trajectories were developed in a study of adaptive farmers in the GTA (Brunet 2006). From that study, it is clear that these farms have distinct business paths and characteristics mainly influenced by land-use restrictions and operator ages. Farmland preservation played a crucial role in farm-level innovation since it decreased uncertainty and stimulated adaptation. Following are brief descriptions of these categories.

Hobby Farmers

Hobby farmers are in their own category since generally they are not concerned with the viability of their operations or their ability to support the farm household.

Leavers

Leavers are divided into two categories: those who leave farming because of retirement and those who leave agriculture for financial reasons. In both cases, the farm might or might not support itself. Commitment to farming varies greatly as well. A common feature is the desire to sell the farm for the highest price, generally for urban development. These farmers might choose to keep running their operations by renting the land for a certain period of time or until development ensues. For them, the decision to sell is the most logical course of action because the rural character of the area as well as the agricultural community has been lost. Farmers often feel outcast or out of place. They fear that too many restrictions will limit their business possibilities in the future. This category is particularly prevalent in urban fringe areas where farmland preservation plans and legislation are non-existent. In fact, 50 percent of informants not in protected zones were considered leavers in this study, stressing the importance of farmland preservation in the peri-urban zone (Brunet 2006). The highly inflated land prices offered by development interests in the GTA seem to convince the most innovative farmers with profitable businesses to sell out. Most farmers were clear in stating that returns from agriculture will never be commensurate with those from selling land for development.

Persisters
In this category, the farm either does not support itself or just breaks even. These farmers rely on off-farm income for survival. They are committed to agriculture and would like to improve the viability of their operations. However, they are also characterized by reduced investments in their operations and limited desires to respond to new opportunities in agriculture. These farmers have very negative outlooks on the future of agriculture, affecting their desire to innovate. Interestingly, all persisters in this study were in protected zones. Most mentioned the future possibility of selling their lands, but this was not imminent. Farming is still a priority. Other reasons for keeping the land include retirement plans, attachment to it, and family values.

Innovators
In this category, the farm might or might not support itself. Farmers have strong commitments to farming and sustainability of the agricultural community. Some even support farmland preservation policies even though they restrict future developments on their lands. In fact, twelve of fifteen farmers in this category were in protected zones, pointing to the possibility of farmland preservation as a way to stimulate creative adaptation in peri-urban agriculture.

These farmers respond to opportunities and constantly search for new marketing and production methods to allow for the continued or future viability of their operations. They either think of succession or have succession plans. They generally have more positive outlooks on the future and are relatively younger than the two other groups. However, some did mention the desire to sell land in the future, but recent farmland preservation legislation has promoted the development of succession plans as speculative land prices are driven down.

CONCLUSION

The expansion of Canadian cities has led to the destruction of some of our best farmland. It is ironic to think that the resource that allowed this expansion is now being destroyed at the expense of its successful development. In response, farmland preservation programs have been initiated throughout the country. The Greenbelt Plan (2005) in the GTA exemplifies this trend. However, simple protection of the land will not ensure the future viability of the near-urban agricultural industry (Beesley 2010; Bunce and Maurer 2005). Many believe that the answer lies in the adaptable nature of agricultural operations, already seen throughout the urban fringe (Zasada 2011). Studies show that agriculture in metropolitan areas has been adjusting for decades in response to the many forces of urbanization (e.g., Bryant and Johnston 1992). Some farmers have decided to capitalize on new opportunities and develop innovative agricultural models. These entrepreneurs will be the next generation of *fringe* farmers in this dynamic environment. Adaptive strategies include alternative production methods, such as community-supported agriculture, and alternative marketing methods, such as direct sales to restaurants. This chapter has demonstrated that the main objective of these strategies is the generation of sufficient income for operations to remain viable and productive while improving links with non-farm rural residents and urban communities.

REFERENCES

AFT (American Farmland Trust). 1997. *Saving American Farmland: What Works.* Northampton, MA: American Farmland Trust.

———. 2008. "Agricultural District Fact Sheet." http://www.farmlandinfo.org/documents/37067/ag_districts_05-2008.pdf.

Beauchesne, A., and C.R. Bryant. 1998. "Agriculture and Innovation in the Urban Fringe: The Case of Organic Farming in Quebec, Canada." *Tijdschrift voor Economische en Sociale Geografie* 90 (3): 320–28.

Beesley, K.B. 2010. *The Rural-Urban Fringe in Canada: Conflict and Controversy.* Brandon: Brandon University.

Bowler, I.R., C.R. Bryant, and M.D. Nellis. 1991. *Agriculture and Environment.* Vol. 1 of *Contemporary Rural Systems in Transition.* London: CAB International.

Brown, H.J., R.S. Phillips, and N.A. Robert. 1981. "Land Markets at the Urban Fringe." *Journal of the American Planning Association* 47: 131–44.

Brunet, N. 2006. "Agricultural Adaptation in the Greater Toronto Area: Farmer Perceptions and Farm Level Trends." MSc thesis, University of Guelph.

Bryant, C.R. 1989. "Entrepreneurs in the Rural Environment." *Journal of Rural Studies* 5 (4): 337–48.

Bryant, C., G. Chahine, K. Delusca, O. Daouda, M. Doyon, and B. Singh. 2009. "Adapting to Environmental and Urbanisation Stressors: Farmer and Local Actor Innovation in Urban and Peri-Urban Areas in Canada." Paper presented at the Innovation and Sustainable Development in Agriculture and Food Conference, Montpellier, France.

Bryant, C.R., and J.A. Fielding. 1980. "Agricultural Change and Farmland Rental in an Urbanizing Environment: Waterloo Region, Southern Ontario." *Cahiers de geographie du Québec* 24 (62): 277–98.

Bryant, C.R., and R.R. Johnston. 1992. *Agriculture in the City's Countryside.* London: Bellhaven Press.

Bryant, C.R., L.H. Russwurm, and S.Y. Wong. 1984. "Agriculture in the Canadian Urban Field: An Appreciation." In *The Pressures of Change in Rural Canada*, edited by M.F. Bunce and M.J. Throughton, 12–23. Geographical Monograph 14. Toronto: York University.

Bunce, M.F. 1985. "Agricultural Land as a Real Estate Commodity: Implications for Farmland Preservation in the North American Fringe." *Landscape Planning* 12: 177–92.

Bunce, M., and J. Maurer. 2005. *Prospects for Agriculture in the Toronto Region.* Toronto: Neptis Foundation.

Chase, R., and S. Hutcheson. 1998. "The Rural/Urban Conflict. Extension Circular." West Lafayette, IN: Purdue University Extension Services.

Daniels, T. 1999. *When City and Country Collide.* Washington, DC: Island Press.

Esseks, D., L. Oberholtzer, K. Clancy, M. Lapping, and A. Zurbrugg. 2009. "Sustaining Agriculture in Urbanizing Counties: Insights from 15 Coordinated Case Studies." Lincoln: Rural Development Program of the National Research Initiative, University of Nebraska.

Fitzsimons, J. 1983. "Urban Growth: Its Impact upon Farming and Rural Communities." *Issues in the Rural Community: Planning* 14: 297–313.

Fitzsimons, J., C.J. Pearson, C. Lawson, and M.J. Hill. 2012. "Evaluation of Land-Use Planning in Greenbelts Based on Intrinsic Characteristics and Stakeholder Values." *Landscape and Urban Planning* 106: 23-34.

Fleury, A., and P. Donadieu. 1997. "De l'agriculture à l'agriculture urbaine." *Le courrier de l'environment* 31: 45-61.

Furuseth, O.J., and J.T. Pierce. 1982. "A Comparative Analysis of Farmland Preservation Programs in North America." *Canadian Geographer* 16: 191-206.

Gasson, R. 1987. "The Nature and Extent of Part Time Farming in England and Wales." *Journal of Agricultural Economics* 38: 167-91.

Hagerstrand, T. 1967. *Innovation Diffusion as a Spatial Process.* Chicago: University of Chicago Press.

Heimlich, R.E., and W.D. Anderson. 2001. *Development at the Urban Fringe and Beyond: Impacts on Agriculture and Rural Land.* Agricultural Economic Report No. 803. Washington, DC: Economic Research Service, U.S. Department of Agriculture.

Hoppe, R.A., and P. Korb. 2000. "The Fate of Farm Operations Facing Development." Draft manuscript. Washington, DC: Economic Research Service, U.S. Department of Agriculture.

Iaquinta, D., and A.W. Drescher. 2000. *Defining Peri-Urban: Towards Guidelines for Understanding Rural Urban Linkages and Their Connection to Institutional Context.* Land Reform 2000/2. Rome: UN Food and Agriculture Organization.

Ilbery, B.W. 1991. "Farm Diversification as an Adjustment Strategy on the Urban Fringe of the West Midlands." *Journal of Rural Studies* 7 (3): 207-15.

Katz, S. 1997. "Farming in the Shadow of Suburbia." *Regional Review* 7 (2): 12-17.

Kormacher, K.S. 2000. "Farmland Preservation and Sustainable Agriculture: Grassroots and Policy Connections." *American Journal of Alternative Agriculture* 15 (1): 37-43.

Lapping, M.B., and M.J. Pfeffer. 1997. "City and Country: Forging New Connections through Agriculture." In *Visions of American Agriculture*, edited by W. Loceretz, 91-104. Ames: Iowa State University Press.

Lynch, L., and J.E. Carpenter. 2003. "Is There Evidence of a Critical Mass in the Mid Atlantic Agricultural Sector between 1949 and 1997?" *Agricultural and Resource Economic Review* 32 (1): 116-28.

Macgregor-Fors, I. 2010. "How to Measure the Urban-Wildland Ecotone: Redefining 'Peri-Urban' Areas." *Ecological Research* 25 (4): 883-87.

Mann, S. 2002. "The Joys of GTA Farming." *Better Farming*, September. http://www.betterfarming.com/archive/cov_jj02.htm.

McInerney, J., M. Turner, and M. Hollingham. 1989. *Diversification in the Use of Farm Resources.* Report No. 232. Exeter: Department of Agricultural Economics, University of Exeter.

Munton, R.J.C. 1990. "Farm Families in Upland Britain: Options, Strategies, and Futures." Paper presented to the Association of American Geographers, Toronto, April.

OFA (Ontario Federation of Agriculture). 2005. "'One Voice' Gaining Ground." Commentary 1005. http://www.ofa.on.ca.

Ontario. Ministry of Municipal Affairs and Housing. 2005. *The Greenbelt Plan.* Toronto: Queen's Park.

Peterson, G. 1982. "Methods for Retaining Agriculture Land in the Urban Fringe in the U.S.A." *Landscape Planning* 9: 271-78.

PlanScape. 2003. *Greater Toronto Area: Agricultural Profile Update Executive Summary.* http://www.planscape.ca.

Roberts, W. 2005. "How Bok Choy Can Beat Sprawl?" *NOW* 17 February. http://www.nowtoronto.com/issues/2005-02-17/news_story.php.

Romero, C., and T. Rehman. 2003. *Multiple Criteria Analysis for Agricultural Decisions*. 2nd ed. London: Elsevier.

Scarfo, R.A. 1990. *A Report of Current Trends and Future Viability of Farming in Maryland's Metropolitan Fringe*. Maryland: Maryland Office of Planning.

Sharp, J.S., and M.B. Smith. 2004. "Farm Operator Adjustments and Neighboring at the Rural-Urban Interface." *Journal of Sustainable Agriculture* 23 (4): 111-31.

Shucksmith, D.M., J. Bryden, P. Rosenthall, C. Short, and D.M. Winter. 1989. "Pluriactivity, Farm Structures, and Rural Change." *Journal of Agricultural Economics* 49: 345-60.

Sinclair, R. 1967. "Von Thunen and Urban Sprawl." *Annals of the Association of American Geographers* 57: 72-87.

Slee, R.W. 1987. *Alternative Farm Enterprises*. Ipswich: Farming Press.

Smithers, J., and P. Johnson. 2004. "The Dynamics of Family Farming in North Huron County. Part 1. Development Trajectories." *Canadian Geographer* 48 (2): 191-208.

Stoneman, D. 2005. "A Tale of Four Farms: Making the Most of Local Market Opportunities to Turn a Profit in a Tough Environment." *Better Farming*, January.

WAC (Watershed Agricultural Council). 2005. WAC Website. http://www.nycwatershed.org.

Winter, M. 2003. "Embeddedness, the New Food Economy, and Defensive Localism." *Journal of Rural Studies* 19 (1): 23-32.

Worden, E.C. 2004. "Growers' Perspectives in Community Supported Agriculture." *HortTechnology* 14 (3): 322-25.

Zasada, I. 2011. "Multifunctional Peri-Urban Agriculture: A Review of Societal Demands and the Provision of Goods and Services by Farming." *Land Use Policy* 28: 639-48.

CHAPTER TEN

Ontario Farmland Trust: Bringing Permanence to Farmland Protection

MATT SETZKORN

GROWING CONCERN FOR FARMLAND PROTECTION IN ONTARIO

The story of farmland preservation in Ontario has been one of incremental progress over several decades as concerns about the far-reaching impacts of farmland loss have steadily grown among farmers, governments, land conservationists, and the public. Some of the first actions toward effective farmland preservation arose from the leadership and collaboration of farmers and local governments in areas such as Waterloo Region and Huron County. It was there that the first municipal farmland protection policies took root. The Foodland Guidelines of the 1970s brought more attention to planning for agriculture at the provincial level, and the introduction of the Provincial Policy Statement (PPS) in the 1990s set the stage for more integrated and comprehensive land-use planning in municipalities across Ontario.

Today, as the province nears completion of an extensive three-year consultation process on the fourth version of the PPS, it is clear that farmland protection will continue to receive careful attention, with additional refinement and improvement of these policy guidelines that strengthen protections for agriculture across the province. The PPS identifies farmland preservation as a core provincial interest and ensures that local and regional planning "be consistent with" PPS policy guidelines to avoid the loss of farmland resources and mitigate any land-use conflicts with agriculture.

In rapidly urbanizing places, such as the Greater Golden Horseshoe that surrounds Toronto, additional provincial legislation, such as the 2005

Greenbelt Act and Places to Grow Act, encourages an integrated, regional approach to protecting key agricultural and natural landscapes, managing population growth, and planning for infrastructure in a way that reduces urban sprawl, makes more efficient use of existing and designated urban lands (infill development, higher-density housing, etc.), and improves the cost effectiveness of municipal service delivery.

Limitations of Policy

Despite many improvements to farmland protection policy over time, Ontario continues to lose tens of thousands of acres of productive land every year. Urban sprawl, scattered non-farm developments in rural and agricultural areas, and development of new aggregate pits and quarries erode Ontario's limited farmland base, with serious implications for the long-term viability of the farm sector.

Land-use policies are always subject to change, and land-use decisions are made again and again to the detriment of agriculture. Population-growth pressures are used to justify the expansion of urban areas at the expense of adjacent farming communities. Mining of aggregate mineral resources is explicitly prioritized by the province over all other land uses, including agriculture, permitting prime farming soil to be stripped away with little commitment to rehabilitation for farm purposes. Non-farm recreational, institutional, and residential uses are introduced to the countryside, often in direct conflict with agriculture.

Land-use planning and development decisions are often political. Farmland preservation policy is only as good as the long-term political commitment to implementation and a strong understanding of the rationale behind planning protections for agriculture. With fewer farmers in Ontario and an increasingly urban population, leadership in government, historically having strong farm representation, has shifted away from a familiarity with agriculture. With this dynamic comes a growing risk of land-use decisions being made without the input of farm communities. With little awareness of the unique needs of the sector and the far-reaching impacts of certain land-use changes in farming areas, seemingly small decisions made by elected officials or disconnected policy makers can result in adverse effects on farm livelihoods and sustainability of the larger agricultural industry in a community or region.

Moving beyond Policy and Responding to Mounting Public Concern
Land-use policy must continue to provide the foundation for farmland protection in Ontario. Policy has clear limitations, however, and has not been effective at *permanently* protecting farmland resources and farm communities across the province as a stand-alone approach. It requires supporting tools and consistent community engagement to ensure that the interests of farmers and the public are effectively reflected in policy, consistently upheld in land-use decision making, and permanently protected.

Ontario farmers and farm organizations recognize that more needs to be done, and a growing number of urban and rural Ontario residents are coming alongside the farm community to push for greater protection of farmland. Greater public interest in food and farm issues has been sparked in part by a burgeoning local food movement in Ontario over the past five to seven years as well as groundbreaking opposition to a recent 2,300-acre limestone quarry proposed on prime farmland. Here, over several years, farmers and urban Ontarians stood side by side in an unprecedented lobbying effort, encouraging the Ontario government to reject the quarry plans. Events organized to oppose the "mega-quarry" drew as many as 30,000 people, demonstrating diverse but united support for farmland protection and agriculture in the face of such a threat. The quarry proponent withdrew its application in 2012, and public interest in the issue has morphed into a "Food and Water First" campaign that advocates for additional policy protection for farmland. The local food movement and the mega-quarry opposition in Ontario have been recent catalysts to building greater public awareness of, and support for, more action on farmland preservation.

It is in this context of evolving land-use policy and increasing public concern about farmland loss that the Ontario Farmland Trust (OFT) has pursued its work since 2004, introducing new methods of community engagement, education, and direct land protection that strengthen and advance farmland preservation in Ontario. The OFT leverages growing public interest in farmland protection to inform policy development and support individuals and communities that want to go beyond policy to permanently protect farmland.

FARMLAND AS A STRATEGIC—AND DIMINISHING—RESOURCE IN SOUTHERN ONTARIO

Despite the vast size of Canada, only 11 percent of the land base has any capability for agricultural production because of limitations of soil, topography, or climate. When we discuss farmland in Canada, it is important to distinguish the different classifications of land according to the Canada Land Inventory (CLI). The CLI ranks the capability of land to support agriculture based upon limitations imposed by the soil on mechanized agricultural activities. The CLI includes seven soil classes. Class 1 soil has no significant limitations on use for crops, whereas Class 7 soil has no capability to support crop or livestock production. "Prime" farmland refers to agricultural land Classes 1–3, capable of producing a wide range of agricultural commodities. Only 0.5 percent of Canada's land area is Class 1 farmland, our most dependable and productive land resource, and only 5 percent of Canada's land falls into the broader prime farmland category (FPRP 2004).

Containing over half of Canada's Class 1 soil, Ontario's farmland is a strategic resource and the single most important agricultural resource in Canada (FPRP 2004). Combined with southern Ontario's moderate climate, this land can produce a greater diversity of crops than anywhere else in Canada—over 200 different commodities, including tender fruit and vegetable crops, grains, and oilseeds (OMRI 2011). This land and its diversity of production provide the foundation for Ontario's agricultural and agrifood industries that provide 700,000 jobs and contribute $34 billion to the provincial economy annually (OMRI 2011).

Yet farmland area in Ontario totals only 12.67 million acres, less than 5 percent of Ontario's entire land area (Statistics Canada 2011). The vast majority of this land is found in southern Ontario, for the northern Ontario landscape is largely rock and not suitable for farming. Southern Ontario is also home to over one-third of Canada's population, and ongoing population growth and urbanization fuel the conversion of much of the country's best agricultural land to non-farming uses. One-third of Canada's Class 1 farmland can be seen from the top of the CN Tower in downtown Toronto, and a large portion of this land is now covered by houses, factories, and highways (FPRP 2004).

Farmland Loss in Ontario: Patterns and Projections

Collecting consistent and accurate data on farmland loss is a challenge; however, it is clear that Ontario farmland is a resource that has been in decline for decades. The *Census of Agriculture* data are the main source of information on the area of land being farmed, but these data have limitations. The census only gathers information on currently operating farms, so when a farm ceases to exist, economically speaking, the land area that it occupies simply vanishes from the census. It is therefore difficult to determine whether that farm was urbanized, abandoned and reverted to forest, or sold to a rural non-farm landowner. Regardless of this challenge, census data on farmland in Ontario indicate that the amount of farmland being actively farmed has been steadily decreasing since the 1950s (FPRP 2004).

The loss of farmland has been quite variable across the different regions of Ontario. The heaviest losses have occurred in northern Ontario and, to a slightly lesser extent, eastern Ontario, where many marginal farmlands could not sustain viable agricultural businesses and were left to naturalize. In southwestern Ontario, there has been much less loss of farmland, with farms largely being maintained for agricultural production. In central Ontario, particularly the region surrounding the Greater Toronto Area (GTA), farmland loss has been over 50 percent. Here the loss is almost entirely the result of urbanization (FPRP 2004). Interestingly, since 1976, overall farmland area in crop production in Ontario has remained relatively consistent, whereas pastureland, on-farm woodlots, and wetlands have decreased substantially (Weersink et al. 2013). We know that expansion of urban areas in southern Ontario typically occurs on prime farmland, for lack of alternatives. The census trends, therefore, might suggest a shift toward more marginal farmlands being brought under cultivation as more productive lands are urbanized and no longer available for farming purposes. Not only the quantity of Ontario's farmland resources lost year over year but also the class of land and its ability to support long-term agricultural production are concerning. Improved data collection and additional analysis are needed to measure the extent to which the most productive farmlands in Ontario are being lost and the permanence of such losses.

The census data show that, between 1976 and 2011, 2.8 million acres of farmland were no longer being farmed because of either urbanization or naturalization of marginal farmland and are unlikely to be returned to agriculture. This is the amount of land required to feed the entire population of Toronto. The *2011 Census of Agriculture* data indicate that we continue to lose farmland in Ontario.

The Ontario Farmland Trust released a study in 2012 that considered *Farmland Requirements for Ontario's Growing Population to 2036*. The research estimated the farmland required to feed Ontario's population within a twenty-five-year planning horizon based upon various population projections, Ontarians' food needs, current farm productivity, and available farmland in Ontario. The study found that, in all low-, medium-, and high-population growth scenarios, Ontario has the potential to lose food self-sufficiency within the next twenty-five years. This potential further raises the urgency of protecting Ontario's farmland resources and questions the extent to which we collectively value and support the ability to be food self-sufficient and to maintain a vibrant farm economy in the province.

DEVELOPMENT OF THE ONTARIO FARMLAND TRUST
Land Trusts and Grassroots Land Conservation in Ontario

Prior to introduction of the Foodland Guidelines under the Planning Act in 1978, a number of grassroots groups formed to encourage the province to adopt measures to permanently protect farmland using conservation easements in addition to improved land-use policy. This included the Preservation of Agricultural Land Society (PALS) in the Niagara Region and the Green Door Alliance in the Pickering area; both continue to pursue some advocacy work.

During the 1980s and 1990s, however, there was a much greater focus on the protection of Ontario's natural heritage, and many new groups were created to fight the urbanization of their favourite woodlands or wetlands. With persistence from organizations such as the Federation of Ontario Naturalists (now Ontario Nature, one of Ontario's leading land trusts), improvements were made to watershed planning and stormwater management, and revisions were made to the Planning Act that strengthened natural heritage policies to include protection of valley lands, rare species' habitat, woodlands, and other natural features.

Community-based land trusts and conservancies also emerged at this time as citizens' groups, working at the local or regional level, dedicated specifically to land conservation. The work of these groups was almost always in the form of preserving natural habitats. At the same time, the farm community developed an approach to environmental protection through the Environmental Farm Plan program, a remarkably successful effort to encourage individual farmers to manage environmental issues on their own farms.

Since the 1970s, the permanent protection of farmland was virtually ignored as a public issue by both conservation and farm groups, particularly in the landscape around the GTA, where land continued to be rapidly lost to urban growth and development. The few citizens' groups that promoted farmland preservation were small voices in the wilderness of urban growth and development. It was from this history of grassroots land conservation in Ontario, and the growing need for a stronger voice for the protection of farmland, that discussions about creating an Ontario-wide farmland trust emerged.

Formation of the Ontario Farmland Trust

The Ontario Farmland Trust began to take shape at a workshop on farmland preservation hosted by the University of Guelph's Centre for Land and Water Stewardship in 2002. Farm and conservation groups came together to discuss concerns about farmland loss in Ontario and the possibility of creating a new organization that would actively promote preservation of farmland in the province. Strong agreement emerged that an innovative land trust model should be developed to build bridges between land conservation and agriculture and have a role in direct land protection, research, and education.

Much of the next eighteen months saw conservationists and farmers interacting to develop the organization's structure and reaching out to Ontario's major farm organizations, largely unfamiliar with land trusts. Initially, there was some resistance to the idea of a land trust among farmers since there were concerns about additional layers of regulation of farm activities. At the same time, however, the provincial government was in the process of releasing dramatic proposals related to agricultural land-use planning, including the Greenbelt Act, changes to the Planning Act, and growth management planning in the GTA—all intended, in part, to protect farmland. The farm organizations suddenly saw land-use planning for farmland preservation as a major public issue and saw value in supporting a new farmland trust that would bring research knowledge and additional insight to good planning decisions for agriculture as well as provide new tools to allow farmers to voluntarily protect their lands.

In 2004, the OFT became a federally incorporated not-for-profit organization and gained charitable status. From the start, its Board of Directors has included representation from Ontario's major farm organizations, the Ontario Federation of Agriculture, and the Christian Farmers Federation of Ontario, as well as academics, researchers, planners, and other land conservation advocates. The OFT has maintained that five of the fifteen board members must represent the interests of farmers, actively receiving nominations for board members from farm organizations. This ensures that the work of the OFT remains supportive of the farm community and advances the long-term viability of Ontario agriculture.

The constitution of the OFT aims to

1. protect and preserve farmlands and associated agricultural, natural, and cultural features in the countryside for the benefit of current and future generations;
2. acquire, secure, manage, and otherwise deal with farmlands, interests in lands, and associated agricultural, natural, and cultural features;
3. research and educate about the value, management, use, and protection of farmlands and associated agricultural, natural, and cultural features;
4. receive, manage, and disburse funds, donations, and bequests; and
5. foster cooperation with individuals, organizations, agencies, and others with similar aims.

Adhering to the Canadian Land Trust Alliance Standards and Practices, the OFT has developed a reputation for fostering collaboration across sectors, providing sound policy advice, and taking direct action to support land conservation initiatives in communities across Ontario. Today the OFT remains the only organization with a province-wide mandate to defend, protect, and preserve Ontario's best farmlands.

THE OFT APPROACH TO FARMLAND PRESERVATION

The OFT focuses its programs and services on three main activities:

1. direct land protection (land securement) through conservation easements and land donations;
2. policy development for improved farmland protection; and
3. research and education to advance land protection and related policy development.

Presenting Options for Permanent Land Protection

The OFT's Land Securement Program is the foundation of the organization's work, permanently protecting farmland by working directly with farmers and other rural landowners to ensure that their lands remain available for agricultural and conservation purposes in the long term. Since 2009, through collaboration with farm owners, conservation agencies, funders, and government partners, the OFT has permanently protected twelve farms and over 1,200 acres using farm easements registered on land titles. Success builds on success, and as awareness of the securement program has spread across the province the OFT has received far more interest than it can currently accommodate. It has developed a rigorous land securement strategy, criteria, and procedures to evaluate each land protection opportunity presented. Working entirely on a donation basis, negotiations to protect twenty-five more farms are now in progress.

The OFT is the only land trust in Ontario that offers the unique service of preserving the agricultural value of farm properties and protecting natural features. The voluntary and collaborative approach to land stewardship and conservation provided by the OFT presents a refreshing and empowering alternative to farmers and other landowners who feel strongly connected to the land and do not trust Ontario's land-use planning system to protect their farms in the future. By donating land to the OFT or working with the trust to establish conservation easement agreements on their properties, landowners can have a direct say in the future uses of their lands and be assured that their farms will be protected from future housing developments, aggregate pits and quarries, dumps, and other non-agricultural uses.

The most common means of direct land protection that the OFT pursues is the farmland conservation easement agreement or agricultural easement. This legal agreement is negotiated between the land trust and the landowner, outlines permitted and restricted uses on the property, and ensures that the landowner's long-term conservation wishes are upheld in perpetuity. The final easement agreement is registered on the property title, preventing future landowners from developing the site for any use other than agriculture and conservation. The OFT becomes the easement "holder" and is responsible for maintaining a relationship with future landowners, monitoring the property annually for easement conformity, and enforcing easement violations if necessary. A charitable tax receipt can be offered to the landowner for the value of the easement, determined by a certified appraiser and "donated" to the OFT. Land or easement donations might be eligible for enhanced tax benefits through Environment Canada's Ecological Gifts program.

The OFT has not actively sought to purchase farmland, but land can be donated to the trust outright or as a bequest. Expecting such a scenario in the future, the OFT is developing partnerships and expertise that will enable land to be leased out at affordable rates to new farmers, alongside farm business training programs. New farmers often have trouble purchasing productive farmland in Ontario to start their agricultural enterprises because of its increasingly prohibitive cost. They also need long-term, secure lease arrangements to be successful, which the OFT can offer. Here it would not only protect farmland but also support the next generation of Ontario farmers.

The OFT, as a provincial land trust, chooses to pursue securement projects that break new ground and demonstrate leadership, innovation in local partnerships, and creative use of conservation easements to help inform the work of other land trusts and broader land conservation initiatives across Ontario.

Enhancing and Strengthening Good Public Policy

Since the start, the OFT has been engaged in supporting federal, provincial, and regional planning and policy development, encouraging greater protections for agriculture and farmland. This involves working alongside other farm and conservation groups to provide input and recommendations on major land-use planning, conservation, and tax policies, including the Provincial Policy Statement, the Greenbelt Act, and the Conservation Land Act, among others. The result has been very positive, with a number of progressive changes made to policy with feedback from the OFT and other groups or policy now being improved during reviews. One notable success was an amendment to the Conservation Land Act in 2006 permitting agricultural conservation easements in Ontario.

The OFT has also put considerable effort into bridging its direct land securement interests and policy development. The trust prioritizes land protection projects that reinforce public policy and good planning for agriculture and conservation. A number of farms protected by OFT easements, for example, are located within or adjacent to the Greenbelt and Oak Ridges Moraine policy areas, where nearby non-farm uses threaten to undermine the intent of this policy to preserve farming, natural areas, and groundwater function. OFT easements demonstrate individual landowner and community commitment to upholding the principles of this policy and thereby reinforce and bring permanence to its intent to protect the area's agriculture and natural heritage.

Another groundbreaking example is a unique partnership developed between the OFT and Simcoe County in 2012 that saw 350 acres of the county's "surplus" land protected by OFT easements. Plans for a landfill development were cancelled because of concerns about impacts on water quality. The OFT worked with the county to craft easements that reinforced local agricultural protection bylaws, with additional protections for water quality, and the land was sold back to farmers, bringing resolution to a once contentious issue in the community.

Municipal partnerships are a key focus for the OFT in both policy and land securement. The trust supports the development of municipal and other government programs and incentives that encourage and invest in the permanent protection of both farmlands and natural heritage lands where most threatened. Several municipalities in Ontario now have cost-share funding opportunities available to them. Land-use policy and easements have limited impacts as stand-alone tools, but when a combination of tools and approaches is used farmland protection efforts can be substantially strengthened and leave lasting impacts.

The OFT will continue to monitor farmland protection policy development and implementation across the province at various levels of government, tracking improvements to policy, policy gaps and needs, emerging trends, and new challenges to achieving effective farmland preservation.

Supporting Research and Education for Farmland Preservation

The OFT's research, education, and outreach activities are closely integrated with its primary land securement and policy objectives, building support for farmland preservation across the province. The OFT produces educational resources, runs farmland protection campaigns, and holds regular events to raise awareness of farmland issues. A long-standing partnership with the Ontario Agricultural College at the University of Guelph supports student learning and farmland research. Annual Farmland Forums bring together diverse stakeholders to discuss emerging trends, ongoing challenges, and new opportunities for farmland preservation. Workshops with farm advisers raise awareness of farmland protection tools available to landowners. Best practices in farmland securement and effective use of farmland conservation easements are shared with other land trusts and conservation groups in Ontario and across Canada.

THE FUTURE OF FARMLAND PRESERVATION IN ONTARIO

Farmland preservation in Ontario will continue to be a collaborative process in the future and require leadership on all fronts: policy, permanent land protection, and authentic community engagement. As long as population growth and urbanization put pressure for development on the rural landscape, farmland loss will continue, to the detriment of farm livelihoods, Ontario's food self-sufficiency, and the overall well-being of communities across the province, both rural and urban. The ongoing loss of tens of thousands of acres of the most productive farmland every year undermines local and provincial economies by removing a perpetual resource capable of producing food forever and sustaining Ontario's largest economic driver: the agri-food sector. We need to recognize farmland in Ontario as the strategic resource it is, ensure that all efforts are made to design policy and programs to mitigate the loss of this valuable resource, and support the long-term viability of farm communities. Farm organizations, land conservation groups, governments, farmers, rural landowners, environmentalists, and food interest groups must see themselves as partners and work together to permanently protect Ontario farmland for current and future generations.

The OFT celebrated its tenth anniversary in 2014, and it continues to look forward to building on the great momentum gained toward improved farmland protection over the past several years through new partnerships, mobilized community support for local food and farmland, and layering of policy and permanent land protection tools that effectively advance land preservation for the benefit of all Ontarians. The OFT is maturing as an organization with a credible voice in policy development and a commitment to on-the-ground action. Over the next five to ten years, the OFT will focus on building organizational capacity, fostering long-term partnerships, enhancing the strategic nature of its land securement program, and expanding its impacts to achieve a vision for permanent farmland protection and a sustainable future for agriculture in Ontario.

The Ontario Farmland Trust vision is a future in which the best farmland in Ontario is valued and permanently protected through sound policy, partnerships, and proactive community engagement; in which diverse farming communities thrive; and in which the protection of farmland, agriculture, and local food production is recognized as the foundation of a sustainable rural economy.

REFERENCES

FPRP (Farmland Preservation Research Project). 2004. *Farmland in Ontario: Are We Losing a Valuable Resource?* Guelph: Centre for Land and Water Stewardship, University of Guelph. http://ontariofarmlandtrust.ca/wp-content/uploads/2013/12/OFT-Farmland-Loss-Factsheet.pdf.

OFT (Ontario Farmland Trust). 2012. *Farmland Requirements for Ontario's Growing Population to 2036.* http://ontariofarmlandtrust.ca/places-to-grow-food/ontario-foodland-to-2036.

OMRI (Ontario Ministry of Research and Innovation). 2011. *Agri-Food Asset Map.* http://www.mri.gov.on.ca/english/publications/documents/Agrifood_Summary_Nov2010_ENG_AODA.pdf.

Statistics Canada. 2011. *2011 Census of Agriculture.* http://www.statcan.gc.ca/ca-ra2011/.

Weersink, A., et al. 2013. *Canada's Supply of Agricultural Land.* Guelph: Department of Food, Agricultural, and Resource Economics, University of Guelph.

CHAPTER ELEVEN

Farmland Preservation Policies in the United States

TOM DANIELS

Farmland preservation involves voluntary programs in which landowners agree to limit the use of their properties to farming and other open space uses in return for cash payments and/or tax benefits. America's experience with farmland preservation is a combination of modest success and inconsistent agricultural policies. The successes—as measured by the number of farmland acres preserved—have been concentrated in a relatively small number of counties, mainly in the northeast and California (Sokolow and Zurbrugg 2003). But nationwide there is a split between the income-oriented agricultural policies of the federal government and the land-use and growth management policies of state and local governments. Even though the federal government has offered a farmland preservation grant program for more than twenty years, local governments largely control land-use planning in America. Getting these governments—townships in the northeast and midwest, counties in the rest of the nation—to coordinate their land-use planning and farmland preservation efforts has often been a frustrating experience. Moreover, targeting federal funds at the most productive agricultural regions has not been fully realized.

Farmers and ranchers own most of America's privately held land, about 915 million acres in 2012 (U.S. Bureau of the Census 2014). The average age of farmland owners is fifty-eight. This means that, within the next two decades, tens of millions of acres will change hands. What the heirs and buyers of those lands decide to do with them will have profound consequences for communities all across the United States.

The *2012 Census of Agriculture* counted 2.1 million farms, with a farm defined as being capable of producing at least $1,000 a year in agricultural commodities. But more than half of all U.S. farms produce less than $10,000 a year (see Figure 11.1). Meanwhile, medium-sized family farms are declining in number, while the number of large commercial farms is increasing. In fact, the top 200,000 farms produce most of America's farm output.

Farms and farmland are not evenly distributed across the United States (see Figure 11.2). If America is divided into four regions, the north-central region with much of the corn belt of the midwest and the wheat belt of the Great Plains has the greatest amount of farmland. But the south has the highest number of farms. The west has California, the nation's leading farm state, and the northeast accounts for only a fraction of farms and farmland. In addition, most of the large farms and ranches are located west of the Mississippi River.

American farmers and ranchers face three main challenges: (1) profitability, (2) passing farms or ranches to the next generation, and (3) resisting the temptation to sell land for development, especially in metropolitan regions where the value of farmland for raising crops and livestock is far less than the value of the land for house lots and commercial sites. Farmland preservation can help farmers and ranchers by providing needed capital to strengthen farm operations, facilitating transfer of the farm or ranch to the next generation, and offering an alternative to selling land for development.

Figure 11.1. Number of U.S. farms and sales and percent of total, 2012.

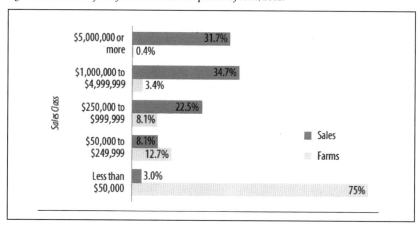

Source: USDA (2012).

186 Farmland Preservation

Figure 11.2: Distribution of acres of land in farms as percent of land area, 2012.

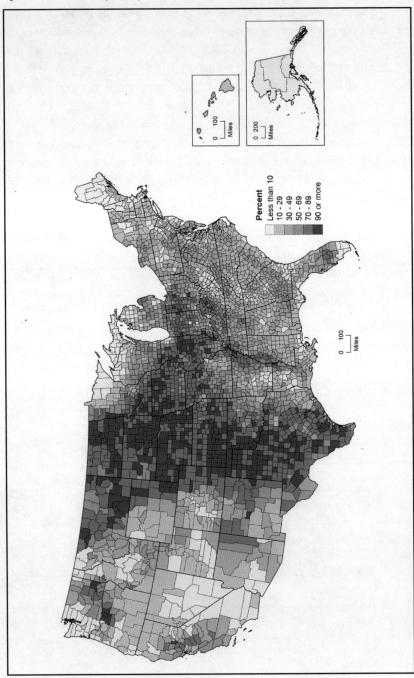

Source: USDA, *2012 Census of Agriculture*, Map 12-M079.

Farmland Preservation Policies in the United States 187

Figure 11.3. Average farm size in acres, 2012.

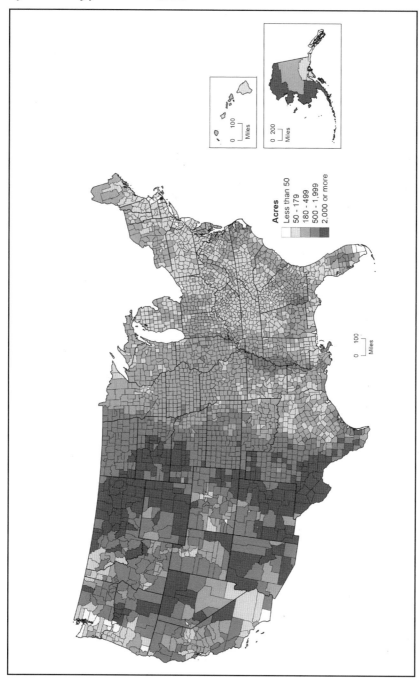

Source: USDA, *2012 Census of Agriculture*, Map 12-M003.

WHAT IS FARMLAND PRESERVATION?

It is important to make the distinction between farmland preservation and farmland protection. Farmland preservation is strictly voluntary, involving the sale or donation of a perpetual conservation easement by a willing landowner to a government agency or qualified private, non-profit land trust. A conservation easement is a legally binding contract, technically known as a deed of easement, that restricts uses of the land to farming and open space. The deed of easement is signed by both the landowner and the government agency or a private, non-profit land trust, and it is recorded at the local county courthouse. The deed of easement runs with the land; if the land is sold or passed on to heirs, then the land-use restrictions continue to apply to all future landowners.

It is possible to overturn a perpetual conservation easement in two ways. First, a government agency can use its power of eminent domain to condemn land under a conservation easement for a public purpose. For instance, if a state highway department needs to construct a public road through preserved farmland, then it can condemn the land, pay the landowner a court-determined sum of money known as "just compensation," take ownership of the land, and build the road. Second, if the government agency or private land trust that holds a deed of easement does not monitor the preserved farmland for compliance with terms of the deed, then the landowner can appeal to a court to have the conservation easement overturned. The holder of an easement has the responsibility to monitor the property—usually visiting it at least once a year—and to enforce terms of the easement. If the easement holder does not perform these duties, then a judge can rule that the holder does not care about the easement and that it is no longer valid. The good news is that more than 25,000 conservation easements have been created in the United States, and fewer than twenty have been overturned (Anella and Wright 2004; Land Trust Alliance 2011).

Farmland protection techniques are not permanent, but they can play an important complementary role in farmland preservation. These techniques include use-value property taxation of farmland, low-density agricultural zoning, urban growth boundaries, right-to-farm laws, agricultural districts, and a governor's executive order to direct state infrastructure projects away from farmland (Daniels and Bowers 1997). These techniques can be changed by an act of a governor, state legislature, or local government. They are political decisions and hence vulnerable to changes in office holders and policy makers.

THE ORIGINS AND GOALS OF FARMLAND PRESERVATION IN AMERICA

Farmland preservation in the United States is over forty years old. The first farmland preservation program arose in Suffolk County, New York (the eastern end of Long Island), in 1974. In 1977, the State of Maryland created the first statewide funding program for the purchase of conservation easements to preserve farmland. Several states and a number of local governments followed. Today farmland preservation programs exist in twenty-six states and more than 150 local governments (www.farmlandinfo.org). More than 3 million acres of farmland have been preserved through the purchase and donation of conservation easements. State and local governments and the federal government have spent more than $4.7 billion to preserve more than 2.55 million acres of farmland (Farmland Information Library 2012). More than 500 private land trusts have listed preserving farmland as an important priority. Although no firm numbers exist for the amount of farmland preserved by land trusts, I estimate that land trusts have preserved about 500,000 acres of farmland. This figure does not include the 1.9 million acres of rangeland preserved by the seven western land trusts that make up the Partnership of Rangeland Trusts (2014).

The goals of farmland preservation vary somewhat from place to place. Yet common goals are

1. to curb sprawling development in the countryside that drives up the price of farmland beyond what farmers can afford, forces up property taxes as new residents demand more services (especially schools), and brings in non-farm residents who complain about the noise, dust, and odor of neighbouring farm operations;

2. to protect high-quality agricultural soils and maintain agriculture as part of the local economy;

3. to manage growth in terms of both location and cost;

4. to maintain the open space and scenic vistas that farming provides; and

5. to support locally grown produce.

Money paid by a government agency or land trust for a conservation easement can help to achieve the above five goals as follows. First, the farmer can choose to sell an easement and thus raise cash without having to sell land for development. The farmer can use the money to set up a retirement fund, reinvest in the farm operation, send children to college, or pay down debts. After the conservation easement is sold, the farmland is restricted to farm use but still has value as farmland. Moreover, the cost of the preserved farmland will be more affordable to other farmers after the easement has been sold. Also, the more farmland preserved in an area, the fewer non-farm residents there are likely to be and thus fewer land-use conflicts. Thus, farmland preservation can stabilize and enhance the business climate for farming and encourage farmers to reinvest in their operations.

Second, highly productive agricultural soils are a valuable natural resource, essential for successful farming. Agriculture is a big industry in the United States. For instance, in 2012 American farmers produced $394 billion in food and fibre (USDA 2012). Farming is a business, not just "a way of life," and it is often an important part of a local economy. Farmers pay local taxes, employ workers, and buy inputs locally. The purchase of a conservation easement puts money in the farmer's pocket, and studies have shown that most farmers use the easement payment to reinvest in the farm operation (American Farmland Trust 1999). Easement payments usually involve a substantial amount of money and thus help with agricultural economic development. For instance, a typical easement payment in Lancaster County, Pennsylvania, is about $3,000 per acre. Thus, the sale of an easement on a 100-acre farm would return $300,000 to the landowner. As more land is preserved over time, ideally, enough farmland can be preserved to maintain a "critical mass" for farming. This critical mass will enable farm support businesses—machinery, feed, hardware, transportation, and processing—to remain profitable and in operation and help to sustain the overall farming industry.

Third, the American Farmland Trust has done many studies on the costs of community services (Farmland Information Library 2012). In every case, the studies show, farmland generates more revenue in property taxes than it demands in public services. Conversely, residential development on average demands more in public services than it generates in property taxes. Hence, farmland preservation is a good fiscal strategy in the long run. In addition, preserving farmland in the right areas can channel development toward areas where it can be serviced by existing infrastructure or little additional investment in infrastructure.

Fourth, the public has little understanding of modern farming. But people do enjoy the open space and scenic views that farming offers. The public often supports preserving farmland for its scenic qualities.

Fifth, food production for local markets can occur with farmland preservation, depending on what the farmers produce. Still, the possibility of fresh fruits, vegetables, and meats for local consumption has become an increasingly popular reason to preserve farmland.

STATE FARMLAND PRESERVATION PROGRAMS

There are two main types of state-level farmland preservation programs. In larger states, such as Pennsylvania and Maryland, the state makes grants to counties that then provide funds to match the grants. In smaller states, such as Vermont, Massachusetts, and Delaware, the state Department of Agriculture preserves farmland directly with the individual landowners.

Successes

A state government has far greater financial resources than a local county or township government. Several states have raised funds through the sale of bonds, and Pennsylvania even adopted a special tax on cigarettes, with the revenues dedicated to the state farmland preservation program.

Pennsylvania leads the nation with more than 460,000 acres of preserved farmland (see Table 11.1) and more than 4,200 preserved farms. This was accomplished over a twenty-three-year period, from 1989 to 2012. In addition, a landowner in Pennsylvania who sells a conservation easement is required to have soil and water conservation plans on the property at the time of sale and to update the plans every ten years. Maryland has used its farmland preservation program as an important element in its Smart Growth effort. By preserving more than 279,000 acres of rural land, the farmland preservation program has reduced sprawl and promoted more compact development. Vermont and New Jersey have preserved 170,000 and nearly 200,000 acres, respectively, but Vermont has concentrated its preservation in its two leading agricultural counties, Addison and Franklin, which have more than 80,000 acres of preserved farmland (Vermont Housing and Conservation Board 2011). In addition, Vermont's Housing and Conservation Board has worked closely with the private, non-profit Vermont Land Trust on many farmland preservation projects. This public-private partnership has enabled more funds to be brought to bear on specific projects and has enabled the state to turn over most of the monitoring of conservation easements to the Vermont Land Trust. Delaware has focused most of its preservation in Kent County, which has nearly 60,000 preserved acres (Goss 2012).

Shortcomings

The effectiveness of state-level farmland preservation programs varies considerably. Some states, such as Maine, New Hampshire, and Rhode Island, have made only token efforts at farmland preservation. Other states, such as Florida and Washington, have preservation programs on the books but have never provided funding to enact them. Major agricultural states in the midwest have been slow to create farmland preservation programs. Both Michigan (1997) and Ohio (1998) have formed programs. Michigan in particular has not adequately funded its program. Michigan has preserved fewer than 20,000 acres, but Ohio has preserved more than 54,000 acres. In both states, the number of applicants and farmland acres offered for preservation greatly exceeds the amount of funding available. California, America's leading agricultural state with more than $43 billion in farm output in 2007, has a small farmland preservation program, rendered largely inactive by the state's budget problems. Even so, California has preserved very little land in its Central Valley, where most of the fruits and vegetables are grown in the United States. Simply put, in most states, farmland preservation has not been made a high priority for public policy.

LOCAL FARMLAND PRESERVATION PROGRAMS

Local farmland preservation programs are most likely to succeed when they are supplemented with state and federal funding sources.

Successes

Leading local farmland preservation programs share a number of key features (see Table 11.2). First, they have agricultural industries worth preserving. At the county level, this usually means an annual value of gross farm output of at least $50 million (Daniels 2004). In contrast, many suburban counties have little farming left, and farmland preservation in these places is geared toward the preservation of open space and some "rural character."

Second, successful counties have adopted agricultural zoning ordinances that allow no more than one dwelling per twenty-five acres. These counties have done careful land-use planning, indicating where

Table 11.1. Leading states in farmland preservation, 2012.

STATE	ACRES PRESERVED (2012)	TOTAL STATE AND LOCAL COST	VALUE OF FARM PRODUCTION (2012)
PENNSYLVANIA	463,595	$1.1 BILLION	$7.4 BILLION
MARYLAND	279,223	$589 MILLION	$2.3 BILLION
NEW JERSEY	199,858	$1.5 BILLION	$1.0 BILLION
VERMONT	170,000	$100 MILLION	$.8 BILLION
DELAWARE	104,960	$187 MILLION	$1.3 BILLION

Sources: Bowers (2012); USDA (2012).

Table 11.2. Leading counties in farmland preservation, 2011.

COUNTY	ACRES PRESERVED	GROWTH BOUNDARY	AGRICULTURAL ZONING
BALTIMORE, MD	54,620	SINGLE	1 HOUSE PER 50 ACRES
LANCASTER, PA	90,280	SEVERAL	1 HOUSE PER 25 ACRES
MARIN, CA	46,640	SINGLE	1 HOUSE PER 60 ACRES
MONTGOMERY, MD	71,865	NONE	1 HOUSE PER 25 ACRES
SONOMA, CA	43,128	SEVERAL	VARIES

Source: Bowers (2011).

development should or should not go. In short, protecting the farmland base has driven the county's overall land-use planning effort.

Third, many of the successful counties have put in place urban growth boundaries to promote more compact development and to limit the extension of sewer and water lines and schools into the countryside. A growth boundary is supposed to contain enough land within it to accommodate projected growth over the next twenty years. Although a boundary can be expanded, farmland just outside it that has been protected through zoning or preserved with a conservation easement helps to reinforce the boundary. Some counties use a single boundary, whereas others use multiple boundaries. Avin and Bayer (2003) identified some 150 growth boundaries and urban service areas in the United States.

Fourth, successful counties have each preserved more than 40,000 acres through the purchase of conservation easements or the transfer of development rights and can preserve more farmland.

Fifth, the land-use planning and farmland protection and preservation techniques comprise a package being replicated in other counties.

Counties that meet the above criteria include Marin and Sonoma Counties in California, Baltimore County in Maryland, and Lancaster County in Pennsylvania. Other counties—such as Chester County, Pennsylvania; Kent County, Delaware; and Addison County, Vermont—have preserved more than 40,000 acres, but zoning in the countryside needs to be tightened up, and not one of these counties has a growth boundary. Montgomery County, Maryland, does not have a growth boundary and instead has used an adequate public facilities ordinance to determine when rural land can be developed at urban densities.

Lexington-Fayette County, Kentucky, is well on its way to joining the five successful counties cited above. It is Kentucky's leading agricultural county, with more than $500 million a year in the production of crops and livestock. The county determined the nation's first growth boundary, known as an urban service district, in 1958. In the late 1990s, an expansion of the urban service area was agreed to but in return for changes in the countryside. The zoning went from one house per ten acres to one house per forty acres, and the city-county government began to purchase development rights from farmers. To date, more than 28,000 acres of Lexington-Fayette County farmland have been preserved, with many more acres slated for preservation (Lexington, Kentucky Government 2014).

As a final note, since 1989, Lancaster County's Agricultural Preserve Board and the private Lancaster Farmland Trust have had a cooperative agreement to coordinate farmland preservation efforts. This public-private cooperation has resulted in a number of jointly funded preservation projects, in particular the preservation of the farm where much of the movie *Witness* was filmed (Daniels 2000). Several counties and states now partner with land trusts to preserve farmland.

Shortcomings

First, land preservation is not a swift process. Procedures for the sale of an easement typically run as follows.

1. There is initial contact with the landowner.

2. The landowner applies to sell a conservation easement.

3. The government agency or land trust ranks applications from several landowners.

4. The government agency or land trust helps the landowner to find a professional appraiser to appraise the value of the

conservation easement. This is a "double appraisal," involving an estimate of the market value of the property (also known as the "before value") and the value of the property subject to the conservation easement (known as the "after value"). The difference between the two values is the value of the conservation easement. Appraisals take time, anywhere from a few weeks to months.

5. If the landowner accepts the offer to purchase the conservation easement, then the government agency or land trust must order a title search. A new boundary survey of the property might have to be ordered to accurately describe the land subject to the easement. This can take up to several weeks. If there are any mortgages on the property, then the mortgage holders must be paid off when the easement is settled, or they must agree to sign a subordination agreement that keeps the easement intact even if they foreclose on the mortgages. If a mortgage holder cannot be paid off at settlement and refuses to sign a subordination agreement, then an easement cannot be executed.

6. At settlement, the landowner receives a cheque for the conservation easement, the parties sign the deed of easement, and it is recorded along with any subordination agreements at the county courthouse.

7. Monitoring and enforcement of the conservation easement begin by the government agency or land trust.

Second, funding is variable. Many farmland preservation programs have long backlogs of applicants interested in selling conservation easements. If funding is not adequate, then some of these applicants might drop out and pursue development options.

Third, purchasing conservation easements can be very expensive. For instance, in Montgomery County, Pennsylvania, just outside Philadelphia, conservation easements have consistently exceeded $10,000 an acre (to as high as $50,000 an acre). Typically, it costs more than $1 million to preserve a sixty-acre farm. In general, when the average price of conservation easements exceeds $5,000 an acre, local governments are hard pressed to fund easement purchases.

Fourth, some local programs and land trusts lack preservation strategies. They simply attempt to preserve whatever land is offered to preserve. Public policies, such as a comprehensive plan and agricultural zoning, should be in place to indicate where farmland should be preserved over the long run.

FEDERAL FARMLAND PRESERVATION

In recent years, U.S. farmland has been converted to other uses at a rate of about 800,000 acres a year. About half of this land is prime farmland, and most of it is in metropolitan counties, where four of five Americans live (NRCS 2009). These metro counties produce about one-fourth of America's food supply and the majority of its fruits and vegetables.

The federal government does not have a coherent strategy to protect farmland (U.S. GAO 2000). Federal farm policy, until the 2014 Farm Bill, had been dominated by farm-income policies that featured direct payments to farmers for the production of corn, soybeans, wheat, cotton, and other crops.

There is no federal farmland policy that states and local governments are required to follow. In Britain, in comparison, there has been a national policy to discourage farmland conversion since the Town and Country Planning Act of 1947. Instead, the federal government has left land-use matters to the control of states, counties, and municipalities. Even so, the government does influence land use and the cost and location of private development through legal rulings by the Supreme Court, tax policy, and more than ninety spending programs. For instance, federal highway projects, federal grants to local governments for sewer and water projects, and the annual mortgage interest deduction for homeowners have subsidized the conversion of millions of acres of farmland over the past sixty years (Daniels and Bowers 1997, 76).

Successes

The federal effort to provide funding for state and local governments and private land trusts to preserve farmland began modestly in 1990 with a program to lend money to states for such preservation. In the 1996 Farm Bill, Congress abandoned this approach and provided $35 million in federal grants to states and local governments with farmland preservation programs. It was hoped that new state and local programs would also start up to take advantage of the federal money. The Farm and Ranchland Protection Policy Act (FRPPA) included in the 2002 Farm Bill was a major funding breakthrough for farmland preservation (American Farmland Trust 2002). The act authorized

$985 million over ten years in federal grants to state and local governments and private land trusts for the purchase of conservation easements to farmland. In the 2008 Farm Bill, Congress authorized $743 million over five years for farmland preservation. The 2014 Farm Bill combined funding for farmland preservation with the preservation of wetlands and grasslands into the Agricultural Conservation Easement Program, with a total of more than $550 million available through fiscal 2018.

Since 1996, the federal government has allocated more than $1.1 billion in farmland preservation grants to state and local governments and land trusts. This money has enabled the preservation of 1.1 million acres (NRCS 2012). Federal funding has become increasingly important in light of the budget cuts that many state and local government programs suffered as a result of the recession of 2007–09 and the slow economic recovery.

Shortcomings
The Natural Resources Conservation Service (NRCS), which administers the federal farmland preservation program, has been criticized for its lack of a preservation strategy. The NRCS has spread money around to every state and made many grants to private land trusts. Spreading the money geographically can win the FRPPA supporters in Congress for future funding. But some states, such as New Hampshire and Rhode Island, have relatively little farming left. Private land trusts tend to operate outside public land-use planning, which determines where land should be developed or preserved. This opens up the likelihood of a lack of consistency and accountability in preservation efforts.

CONCLUSION

In areas that have strong agricultural industries, farmland preservation can play an important role in ensuring the future of farming. Land-use planning in America has traditionally meant "planning for development." Now many communities are recognizing the need to plan for the preservation of land as well.

In America's metropolitan regions, the value of farmland for farming purposes is less than the value of that land for house lots, strip malls, and office parks. Local governments in metro regions that attempt to rely solely on the purchase and donation of conservation easements will be hard pressed to find the money to pay high prices per acre for easements or to create large contiguous blocks of preserved farmland. The risk is that these local governments will simply "throw money" at the farmland problem and preserve only "islands" in a "sea" of development.

Conversely, in very rural areas, the value of a conservation easement is likely to be so low as to discourage farmers from selling or donating their farmlands.

Successful farmland preservation programs combine significant local and state funding for farmland preservation with a package of farmland protection techniques, especially low-density agricultural zoning to minimize non-farm uses in farming areas and urban growth boundaries to limit the extension of central sewer and water lines into the countryside.

The package approach will become more popular over time as greater pressure is placed on farmland in metropolitan counties. America is facing population growth of more than 80 million people between now and 2050, and most of this growth is expected to occur in metropolitan regions. If energy costs continue to rise, importing food from more than 1,000 miles away will be less feasible, and local production will become not just more attractive but also more necessary. But first the farmland base has to be stabilized for the future.

REFERENCES

American Farmland Trust. 1999. *From the Field: What Farmers Have to Say about the Vermont Farmland Conservation Program*. Washington, DC: American Farmland Trust.

———. 2002. "Congress Commits $1 Billion to Farmland Protection Program." *American Farmland*, Spring 2002.

Anella, A., and J.B. Wright. 2004. *Saving the Ranch: Conservation Easement Design in the American West*. Washington, DC: Island Press.

Avin, U., and M. Bayer. 2003. "Rightsizing Urban Growth Boundaries." *Planning* 69 (2): 22–26.

Bowers, D. 2011. *Farmland Preservation Report*, November–December. Street, MD: Bowers Publishing.

———. 2012. *Farmland Preservation Report*, September. Street, MD: Bowers Publishing.

Daniels, T. 2000. "Integrated Working Landscape Protection: The Case of Lancaster County, Pennsylvania." *Society and Natural Resources* 13 (3): 261–71.

———. 2004. "Metropolitan Expansion Policy: Forging the Package." Paper presented at the American Planning Association National Conference, Washington, DC, 27 April.

Daniels, T., and D. Bowers. 1997. *Holding Our Ground: Protecting America's Farms and Farmland*. Washington, DC: Island Press.

Farmland Information Library. 2012. http://www.farmlandinfo.org.

Goss, S. 2012. "Sussex County Farm Preservation Lags behind Rest of State." *Sussex Countian*, 3 July. http://www.sussexcountian.com/news/x1446670644/Sussex-County-farm-preservation-lags-behind-rest-of-state.

Land Trust Alliance. 2011. *2010 National Land Trust Census Report*. http://www.lta.org.

Lexington, Kentucky Government. 2014. "Purchase of Development Rights. . . ." http://www.lexingtonky.gov/index.aspx?page=497.

NRCS (Natural Resources Conservation Service). 2009. *National Resources Inventory, 2007*. Washington, DC: U.S. Department of Agriculture.

———. 2012. *Farm and Ranchlands Protection Program: Program Information by Fiscal Year*. Washington, DC: NRCS. http://www.nrcs.usda.gov/wps/portal/nrcs/main/national/programs/easements/farmranch.

Partnership of Rangeland Trusts. 2014. http://maintaintherange.org/About%20Us.html.

Sokolow, A., and A. Zurbrugg. 2003. *The National Assessment of Agricultural Easement Programs: Profiles and Maps—Report 1*. Washington, DC: American Farmland Trust and University of California at Davis.

United States. Bureau of the Census. 2014 *Statistical Abstract of the United States, 2012*. Washington, DC: U.S. Department of Commerce.

———. Department of Agriculture. 2009. *2007 Census of Agriculture*. Washington, DC: USDA.

———. 2014. *2012 Census of Agriculture*. Washington, DC: USDA.

———. General Accounting Office. 2000. *Community Development: Local Growth Issues—Federal Opportunities and Challenges*. GAO/RCED-00-178. Washington, DC: GAO.

Vermont Housing and Conservation Board. 2011. *Annual Report to the General Assembly, 2011*. Montpelier: Vermont Housing and Conservation Board.

CHAPTER TWELVE

Planning for the Future of Agriculture

BOB WAGNER

As cities in the United States continue to grow, most sprawl out over some of the best farmland. Established where fertile land met transportation routes, such as rivers and train lines, cities such as Atlanta, Chicago, and Los Angeles today include seemingly endless housing and commercial subdivisions linked by a labyrinth of highway networks—with scarcely a productive green acre in sight.

Contrast those megalopolises to a new breed of city headed by progressive planners and residents who view growth as an opportunity to preserve nearby farms and scenic open spaces, riparian and wildlife habitat, and, most important, the local food that they provide. According to studies by the American Farmland Trust (AFT), 86 percent of the fruits and vegetables are grown at the edges of urban areas (American Farmland Trust 2002).

Communities in every region of the country are beginning to take progressive steps toward planning a future for agriculture. "People tend to think farms are not needed in urban areas, so they ignore them until it is too late, relying on the idea that farms elsewhere are enough," said Don Stuart, former director of AFT's Pacific Northwest Regional Office.[1] He added, "Without planning, farmland[s] get so fragmented in urban areas [that] they are subject to increasing conflicts from surrounding non-farmers and are placed under greater pressure" to convert out of agricultural use.

Effective plans include land-use policies and programs to keep land available and affordable for farming, such as the purchase of development rights (PDR), programs, and agricultural districts. They also include economic development tools to make farming profitable, such as direct marketing, value-added processing and agri-tourism, and initiatives to promote conservation practices and the environmental benefits of working landscapes.

"Planning for agriculture is as important as planning for development," declared Jill Schwartz, AFT's former marketing director. "Effective plans help make farming economically viable and environmentally sustainable. And when that happens, communities benefit from the multiple values of farming—jobs for local residents, wildlife habitat, scenic vistas, and community character."

Vibrant farms located at urban edges can infuse local economies with new life. Thriving farmers' markets, bustling farm stands, and popular tourism opportunities such as autumn pumpkin harvest festivals can circulate new dollars in a community and link non-farmers with local sources of food.

"When counties think about planning, they often take a map of the county and start restricting different land uses," stated Gerry Cohn, former director of AFT's Southeast Regional Office. He added, "When we look at the word *planning* in [a] broader sense, we need to think about a vision for the future that includes agricultural economic development as well as land-use policies." Farmers, he disclosed, need to be confident that agriculture will remain a vibrant industry in a community well into the future: "If there's no vision of being profitable on the farmers' part, there's no reason to stay in farming."

SARATOGA COUNTY, NEW YORK

Saratoga County is a picturesque collection of upstate New York towns set among horse farms, dairies, and apple orchards. However, threats to what has been, historically, an agriculturally important county stem from the 1960s, when Interstate 87 was built between New York City and Montreal. "When [the I-87] was completed in the mid-'60s, our little ag county of 80,000 became home to suburbia," Larry Benton, Saratoga County planner, explained.

Since then, the county has lost 130,000 acres of farmland to development to accommodate New York's second-fastest-growing population. County leaders, however, fully aware of the $100 million that agriculture pumps into the local economy annually, and recognizing its importance to the county's quality of life and scenic views, stepped up with a far-reaching plan that ultimately resulted in a county-funded PDR program as well as an agricultural economic development program. The county supervisors' decision to earmark $1 million for PDR was influenced in part by a public opinion poll showing that 79 percent of the county's voters favour PDR.

By 1992, just 13.5 percent of the county remained in farming; by 2000, the county's population had reached 200,000. "That's the pressure," Benton said. "Our reaction has been a county plan calling for preservation of agriculture and directing growth into the Northway corridor."

Since the 1970s, Saratoga has supported agricultural districts, which create a host of protections for farmers, such as favourable tax assessments and protection from "nuisance" complaints. Some of the county's nineteen towns, each with the zoning power of home rule, have exercised their authority to set local land-use policies that preserve farmland and open space. Several of them are considering offering PDR funds that allow farmers to voluntarily sell their development rights, and more than a dozen have adopted "right-to-farm" laws that protect farmers' abilities to conduct day-to-day agricultural activities.

In the 1970s, the county created an "agricultural districts" advisory panel made up of businesspeople, county officials, and farmers, expanding it under a 1992 state law that also opened up grant opportunities for formal agricultural and farmland protection boards. Four years later the panel created an integrated plan to protect agriculture that features recommendations such as

- a county-wide right-to-farm law;
- a county PDR program;
- a public education campaign to raise awareness about the importance of agriculture in Saratoga;
- encouragement to towns to adopt "farmer-friendly" land-use laws, including conservation-oriented development that clusters residential and commercial buildings; and
- innovative economic development opportunities for farmers.

"In many ways, Saratoga County is a good example of a county taking action," stated David Haight, AFT's New York director, who, along with other AFT staffers, has helped county and town leaders with public education, planning and policy measures, and facilitating farmland protection projects.

County officials take the economic development component of their plan seriously. With a neighbouring county, Saratoga jointly hired an agricultural economic development specialist who promotes agriculture and forges economic opportunities for farmers. The specialist created a Farmer-to-Restaurant Networking Day, first held in 2003, to build direct sales potential; collaborated with others to plan the county's popular Sundae on the Farm tourism event, which brings about 2,000 people to a local dairy farm every year; helped the Saratoga Farmers Market Association to revamp its marketing efforts; and applied for a grant to help Saratoga farmers better market their products to New York City restaurants and the popular GreenMarket farmers' market network.

PIERCE COUNTY, WASHINGTON

An urbanizing county snaking between Puget Sound and the Cascade Mountains, Pierce County produces berries, vegetables, and high-value flower bulbs..But with no local fruit-processing centres and a booming commercial sector that has gobbled up farmland, the county's agricultural base is shrinking. Between 1992 and 1997, Pierce County lost 8,000 acres of farmland—or one farm every two weeks.

It's high time to plan for agricultural protection, say farmers and growth management advocates such as Dick Carkner, who chairs the Pierce County Farm Advisory Commission and views neighbouring counties such as King as models of careful planning for the future of agriculture. Carkner, a fruit and vegetable farmer who direct-markets his products to Tacoma residents, is at the forefront of a budding movement to stabilize the county's shrinking agricultural base. "We wish we were protecting farmland in a formal process," Carkner revealed. "A lot of the best farmland is already lost to development, in particular in the river valleys. We're well behind the curve, but we're optimistic."

Pierce County is a community at the cusp of solid land-use planning with agriculture as an important element, or so Carkner and others hope. He points to positive signs such as the county council and executive appointing the advisory commission (and funding it with a modest annual budget), approving a new right-to-farm law, and hosting a Farm-City Forum to bring together farmland protection advocates from western Washington. The Pierce County Farm-City Forum was one of a series that AFT launched to bridge the gap between farmers and their growing number of non-farm neighbours. Such events emphasize how an urbanizing area can benefit from agricultural land preserved nearby and the role that city residents can play in keeping farming viable.

"We were trying to get policy makers, farmers, and city residents to listen to one another," disclosed Cheryl Oullette, former president of the Pierce County Friends of Family Farmers and a local hog farmer who attended the forum. "Farmland loss is a major, major issue; we need to do something if we want these farmers to be here next year."

The forum led to the creation of four task forces charged with working on a variety of issues likely to be at the core of the county's plan for agriculture, such as creating a PDR program, developing programs to help farmers market their products locally, and amending regulations that create barriers to economically viable farming.

Some farmers have found an opportunity in the challenge of saving farmland. Carkner, who farms just two miles from Tacoma, has turned his city-side location into an economic windfall, selling his vegetables through a community-supported agriculture operation (a shareholder program that provides "subscriptions" of weekly produce to residents who "join" the farm) and city farmers' markets. "The farmers who are struggling are those still competing in the wholesale market as opposed to taking advantage of their proximity to these consumers," he said. "Farmers might complain about the traffic, but those are customers, and they buy food. There are a lot of them."

Oullette's group has been working hard to generate more local dollars for farmers and to build better relationships between urbanites and county farms. They established an annual Harvest Fest celebration in the fall, creating a way for families to meet farmers growing local food. In 2002, nearly 5,000 residents visited seven local farms. "Families can learn what's being grown in the area and learn more about the local agricultural economy, while the farmer can take a break and meet his public," she explained. "We're trying to get people to buy local rather than buying from California or Chile."

County leaders have been supportive of saving both the industry and the land on which it depends. The farm advisory commission over the past few years has argued successfully for a host of protection measures, including a conservation assessment of five dollars per parcel that raises funds to support farmland protection and conservation technical assistance for farmers.

CARROLL COUNTY, GEORGIA

Home to more than 1,054 farms, Carroll County supports more agricultural operations than any other county in the state. Its strong beef cattle industry drives a $155-million-a-year agricultural economy made up of a committed farm populace.

In 2001, a group of those farmers approached the Rolling Hills Resource Conservation and Development (RC&D) council to ask about farmland protection measures. The county, located just fifty miles from Atlanta, has experienced considerable population growth: 27 percent between 1980 and 1990 and 22 percent in the following decade. New building permits have followed apace, accelerated by highways that lead directly to Atlanta.

The concerned farmers helped to trigger a movement to better plan for the future of agriculture in Carroll County. Leaders from the farming community, agribusinesses, local officials, and conservation groups gathered in a "stakeholders" meeting to establish consensus on farmland preservation measures.

"Because of our proximity to Atlanta, this is a perfect place for people who want to live in the suburbs and drive into town for work," declared Cindy Haygood, Rolling Hills RC&D coordinator and an employee of the USDA Natural Resources Conservation Service. Adding to the development pressure is Carroll County's wealth of flat, prime land, on which it is easy to build.

The group, the Farmland and Rural Preservation Partnership, set goals—including better public education about the importance of local agriculture, more economic opportunities for farmers, minimum lot sizes in rural areas, a voluntary transfer of development rights program, agricultural districts, and stronger right-to-farm laws—and presented them to the county's Board of Commissioners. New ideas stemming from an intensive two-day farmland protection workshop in the spring of 2003, which several members of the partnership attended, are also being considered. The workshop—coordinated by AFT, the Association of County Commissioners of Georgia, and the Georgia Agribusiness Council—included discussions about the nation's most successful plans for agriculture and provided participants with chances to share ideas about which techniques will work best in Georgia.

The board remains interested in the issue and willing to consider farmland protection measures as part of a county process to update its comprehensive land-use plan. Robert Barr, commission chairman, considers perpetuating agriculture a real priority in the community. "It's an industry—not just part of the heritage," Barr said. "There's a quality-of-life issue, but more than that the industry of agriculture itself in Carroll County is really large." All of this hard work paid off in November 2004 when voters passed, by a 2:1 margin, a measure to extend a Special Local Option Sales Tax of 1 percent. This funding source will generate $3 million for buying easements on farmland, marking the first county PDR program in Georgia. Cohn said that the strength of the farmland partnership lies in its diversity. "The diverse representation can look at the complete farmland protection picture," he stated. "They are determined to find ways to help farmers make a living farming."

In addition to the PDR program, the partnership, with help from a federal grant, has brought a farmers' market to Carroll County. The first season, Haygood revealed, was a roaring success, with a "fair atmosphere" and plenty of spin-off business for Carrollton stores and restaurants. The farmers' market is a prime educational tool to inform the non-farm public about the county's agricultural resources, Barr explained. Agriculture "preserves the one thing that draws people to Carroll County—its rural nature."

MASSACHUSETTS

Agriculture in Massachusetts is changing. Not only is the land area devoted to agriculture continuing to contract, but also the type of agriculture taking place is evolving from one dominated by the production of wholesale products to one increasingly oriented to retail sales.

The *2002 Census of Agriculture* reports that only 10 percent of the state's land area can now be classified as "land in farms," down from 56 percent in 1910. In an increasingly urbanizing state, it is not surprising to find farmers shifting to higher-valued crops and products to offset rising land values and costs of production and with customers "at the ready" producing products for the direct market trade. According to the 2002 census, "nursery, greenhouse, floriculture and sod" now represent the largest single segment of agriculture in the state, accounting for 40 percent of the total market value of agricultural products sold in that year. In contrast, in 1982, "livestock, poultry, and their products" represented half of the value of agricultural products sold. Further indication of the shift in agriculture in the state is the more than 100 percent increase in the "value of agricultural products sold directly to individuals for human consumption" between the 1982 and 2002 census years (USDA 2004).

"The opportunities for farmers in [Massachusetts] to tap into direct-to-consumer markets seem limitless right now," notes former commissioner Douglas P. Gillespie of the state's Department of Agricultural Resources. "With more and more people concerned about where their food comes from, expanding immigrant markets, and the attractiveness of niche products, the demand for locally grown products has never been higher."

Since the late 1970s, Massachusetts has been a leader in the United States in recognizing the importance of protecting farmland for future generations. In 1977, the state created one of the first statewide PDR programs in the country, the Agricultural Preservation Restriction (APR) program. Over 800 farms covering 68,000 acres of land have been enrolled. Through the APR program, valuable farmland has been protected, and much-needed capital has been made available to participating farmers to expand, improve, and diversify their operations. AFT's 1998 analysis of the program, "Investing in the Future of Agriculture: The Massachusetts Farmland Protection Program and the Permanence Syndrome," found that, along with 66 percent of the participants using the proceeds of selling their development rights to retire debts, more than half were also using the proceeds to reinvest in their farms (American Farmland Trust 1998, 20).

But despite the APR program, farmland continues to be lost to urban conversion. Acknowledging that permanently selling development rights is not a choice that every farmer might make, and that significant opportunities exist to retool farming operations to take advantage of new markets and a growing local consumer base, the state Department of Agriculture created a companion to the APR program that focuses on business planning and new business ventures: the Farm Viability Enhancement Program.

Introduced in 1994, the Farm Viability program offers applicants free assistance in developing business plans and new business models for their farms and then the opportunity to apply for one of three grants to assist in implementation of the business plans. The goal of the program is to help improve the bottom lines of participating farms through new or improved management and conservation practices, diversification, direct marketing, value-added initiatives, and/or agri-tourism. In recognition of the need to link stabilization of the land base with economic development, the implementation grants require the acceptance of a protective covenant on the land held by the participating farmers: a five-year covenant for a $20,000 grant and a ten-year covenant for a $40,000 grant or a $60,000 grant for farms greater than 135 acres. As further reinforcement of the commitment of the state to farmland protection, farmers enrolled in the APR program receive priority for business-planning assistance. However, an APR farm is not eligible for the implementation grants.

Since 1996, when funding was first made available, the Farm Viability program has assisted 377 farms with business plans, of which 73 percent have received implementation grants. One of the program's early success stories saw the Cook farm in the Town of Hadley add an ice cream parlour to its dairy operation. Gordon and Beth Cook found the technical support extremely helpful in taking on the new enterprise, and the business plan assisted in their efforts to borrow additional capital for the project. Although expanding marketing opportunities is an outcome of a majority of the projects, many others improve environmental sustainability, and investments can be seen in irrigation systems, manure-handling structures, and updated sprayer equipment.

Since inception of the program, over 37,135 acres of farmland have been covered by the protective covenants required for receipt of the grants. "Planning for agriculture really needs to address both the land-use and the economic needs of farmers. We feel that to the extent that [the] state government can play a role in either area, we have a one-two punch with the APR program and now the Farm Viability program," declared Gillespie.

CONCLUSION

Farmland preservation has been a priority in many areas of the United States. This chapter has highlighted different planning practices, experiences, and approaches to countering farmland loss. For more information on the issues covered in this chapter, contact American Farmland Trust's Farmland Information Center at 800-370-4879 or visit the website at www.farmlandinfo.org.

ACKNOWLEDGEMENTS

A version of this story first appeared in American Farmland Trust's magazine, *American Farmland*, summer 2003 issue.

NOTES

1. This chapter highlights different planning practices, experiences, and approaches to countering farmland loss. These perspectives were gathered from various professionals who work in this area.

REFERENCES

American Farmland Trust. 1998. *Investing in the Future of Agriculture: The Massachusetts Farmland Protection Program and the Permanence Syndrome*. Northhampton, MA: American Farmland Trust.

American Farmland Trust. 2002. *Farming on the Edge: Sprawling Development Threatens America's Best Farmland*. Northhampton, MA: American Farmland Trust.

United States. Department of Agriculture (USDA). 2004. *2002 Census of Agriculture*. https://www.agcensus.usda.gov/Publications/2002/.

CHAPTER THIRTEEN

Farmland Preservation in Australia

TREVOR BUDGE AND ANDREW BUTT

This chapter sets out an overview and analysis of the protection and management of productive farmland in Australia and the policy approaches and regulatory responses at national, regional, and local levels. It focuses on the economic and land-use scene and the particular farmland preservation initiatives taken in Australia in response to pressures on such areas. To provide a contextual setting, the chapter initially examines the issue in terms of Australia's geographical characteristics and its natural resource base. These factors severely constrain the scale and distribution of population and agricultural production in the country. The population distribution pattern in Australia is based upon a small number of large metropolitan areas in coastal regions. Highly productive farmland is found only in limited areas generally at the fringes of the continent and often at the metropolitan edges. Elsewhere, scale of production, access to and management of water, and ongoing agronomic conditions and improvements are consistently central to discussions of agricultural viability, sustainability, and approaches to preserving highly productive farmland.

The respective responsibilities for land-use planning, policy making, and regulatory controls among the various government jurisdictions in Australia comprise the other major contextual issue. These powers are essentially in the hands of the six Australian states, not the federal government, and in turn they are largely delegated for administration and implementation purposes to a myriad of local government authorities. "Australia has six states and two mainland territories. Under the Australian Constitution, the state governments have prime responsibility for land administration and public land management. The Australian Government has a limited land ownership and management role. Its primary role is to promote more efficient land management and land allocation" (Lesslie and Mewett 2013, 5). Historically, the approaches to farmland preservation that have been undertaken by each state jurisdiction have largely been reactive. Policy and strategic land-use planning initiatives to protect productive agricultural land under imminent threat of land-use conversion at the fringes of Australia's five major metropolitan areas have generally been vague and disjointed rather than uniform regulatory approaches that have sought to "lock up" and permanently preserve such areas. More recently, that agenda has become more complex as it has faced concerns regarding the potential impacts of mining operations on productive farmlands and decisions about irrigation for agricultural production versus water allocations for environmental needs. This latter set of issues has occurred in a number of areas that have traditionally been relatively isolated from competing interests over farmland.

The past decade in Australia (and particularly in the period since the preparation of a chapter on Australia in the first edition of this book) has seen only a few shifts in overall government policy approaches, legislation, regulation, or practice at the federal or state level in relation to preserving high-quality productive farmland. However, importantly in the same period, the issue of farmland protection and preservation has become much more widely discussed and debated across the nation than in any recent period. Heightened interest and emerging conflicts might be the precursors of stronger action in response to the rising pressure for more definitive action. This development has resulted from a range of politically contentious challenges that relate to farmland protection: namely, the implications of sustainable water policy, the long-standing drought conditions of most of the 2000s, major mining expansion, and concerns over the continued outward growth of Australia's vast and sprawling cities. Additionally, there has been greater community awareness of and debate on issues of foreign ownership of farmland, food "security" in its many forms, food quality, more locally sourced food, and the availability of farm products, particularly when affected by natural events increasingly linked to climate change.

Across these agendas on farmland preservation, Australia finds itself caught between a series of seemingly incompatible forces and directions. Large multinational corporations have driven an unprecedented mining boom with significant export earnings for the country through energy commodities such as coal, gas, and the future exploitation of coal-seam gas and shale-oil deposits. These interests have begun to collide with popular and farming concerns about food security, including biosecurity, protection of rural farming assets, and overreliance on carbon-based energy sources. Compounding this public policy dilemma and debate is that large areas of Australia, including the irrigation-dependent Murray Darling Basin (which supports 40 percent of Australia's food production), have emerged from a decade-long drought that created water allocation conflicts between food production and riverine environments. Irrigation-based agriculture in this river basin alone accounts for 28 percent of the gross value of Australian agriculture (Australian Bureau of Statistics 2012b), with crops such as grapes, fruits, rice, cotton, and vegetables and dairying dominant. Concurrent with this situation is that record levels of population growth and urban development in Australia's largest metropolitan areas (driven largely by migration) have often occurred at the expense of productive land in peri-urban regions.

In response to these conflicting forces, Australian state governments and some local governments have been active with policy initiatives and land-use planning strategies, rather than prescriptive regulations, to find and balance local solutions where these forces intersect and potentially clash in major agricultural regions. Australia's resource extraction industries, until recently, have generally been located well away from productive farmland, and when closer their scales of operation have been comparatively small. But that convenient occurrence has hit a barrier. The increasing scale of resource extraction, with its consequent actual or potential effects on farmland and underground water supplies, the alleged risks of coal-seam gas "fracking," and the potential social and economic impacts on towns and communities, signal growing discontent across many regions, such that there is now a large-scale revolt across many agricultural heartlands. Two competing resource takers have collided head on in many areas. The mining and resource industries point to benefits from the transformative economic returns to Australia in supplying raw materials to an energy-hungry world. Conversely, primary producers argue that Australia's rural heritage and long-term capacity to supply high-quality food are under attack. This has been further compounded where the limits to confidence in water allocation, particularly in Australia's largest river basin, have been reached, and in many communities environmental forces, in their view, have won the day. This outcome has shaken assumptions of technological solutions to a seemingly unlimited productive capacity across many areas of Australia.

This chapter explores these recent debates and policy responses by providing an overview of the agricultural scene, the basis and rationale for agricultural activity and change in Australia, a review of land-use planning approaches to farmland protection, and a consideration of the emerging resonance of food policy and mining impacts in political and public debates. The chapter concludes with two examples of legislative and policy measures taken by the state governments of Queensland and South Australia and a case study of the Victorian state government's implementation of a metropolitan strategy for Melbourne, Australia's second-largest city, along with some observations on likely future directions.

THE SIGNIFICANCE OF AGRICULTURE IN AUSTRALIA

Agriculture was the foundation industry and land use in Australia from European occupation and expansion in the early nineteenth century. However, the people of the nation that was once termed "the country that rode on the sheep's back" now reside mostly in a small number of large coastal cities (60 percent in the five largest cities). Less than 3 percent of the Australian workforce is now employed in agriculture, and less than 15 percent of Australia's export income comes from agricultural products—down from close to half in the 1970s (ABARES 2012). Table 13.1 shows the value of Australia's major agricultural commodities as export industries. These commodities are now dominated by large-scale agricultural industries mostly in areas where land-use conflicts are minimal. Conflict is much more likely to occur in areas of intensive production for the domestic market, such as in vegetables or smaller-scale family farming operations.

With only 23 million people, Australia's domestic food consumption needs are relatively modest compared with the level of production and the value of exports, which represent about 75 percent of gross farm production. Further, it is estimated that Australia grows enough food for 60 to 80 million people under current practices (Prime Minister's Science, Engineering, and Innovation Council 2010). Preventing the loss of productive farmland is understandably an issue of limited interest to governments generally, partly because of these factors. The dominance of other export products such as coal, iron ore, and other mineral resources, and the much greater importance of international earnings from service industries such

Table 13.1. Value of selected Australian agricultural exports.

COMMODITY EXPORTS	ANNUAL VALUE
GRAINS AND OILSEEDS	11.1 BILLION
COTTON	2.7 BILLION
WINE	1.8 BILLION
HORTICULTURE	1.7 BILLION
MEAT AND LIVESTOCK	6.8 BILLION
WOOL	3.1 BILLION
DAIRY	2.3 BILLION

Source: ABARES (2012); values in Australian dollars.

as tourism and education, have also limited the government's interest in reducing farmland loss.

The size of Australia, the limited history of farming, and the apparent supply of land available to be "developed" for farming purposes have meant that preserving productive farmland has ranked low as a policy priority and certainly as a topic worthy of legislation or regulation.

A chronicle of the relative lack of action by national, state, and local governments in preventing the loss of farmland to urban growth across Australia appears to confirm that view (Budge et al. 2012). Yet the politics of agriculture, and especially of water policy, continue to loom large in political and public debate, even as incremental land loss and localized industry decline are neglected as policy issues. The pressure for change has been such that the Australian government has committed to developing a National Food Plan (Department of Agriculture, Forestry, and Fisheries 2012), though an examination of the matters addressed in that document reveals that the central motivation has not been concern for the loss of productive farmland. The federal government continues to reiterate that land use and its regulation are constitutionally matters for the states.

Since the early 1970s, a significant process of restructuring in Australian agriculture has occurred as a result of the declining terms of trade, loss of preferential markets, and emergence of neo-liberal policy approaches that have reduced direct subsidies and import protections. Australia-wide, farm numbers decreased by approximately 10,000 or 7 percent in the decade to 2009 (Australian Bureau of Statistics 2010). Remaining farm businesses have generally increased in scale to maintain viability in the face of declining terms of trade (ABARES 2007). Comparatively high levels of growth in productivity in the agricultural sector in Australia over several decades

(Mullen 2010) have been both a cause and a consequence of globalization and the restructuring of local industries within agricultural supply chains and new policy environments. The consequences of restructuring have been unevenly felt among industries and across regions. These changes have also had significant effects on rural communities and populations, resulting in population loss and service centralization and decline, often concurrent with more general trends in rural decline as governments and commercial services withdraw from small and medium-sized rural towns.

The trends toward fewer, larger farms, while dominant in most productive agricultural regions, have not been uniform. In fact, in peri-urban areas, small-scale farms have grown in number, with production concentrated in a number of key industries, notably cattle production and intensive animal industries. In the peri-urban regions surrounding Australia's larger cities, clusters of farm numbers and the share of these regions' production in overall national agricultural output have not changed dramatically (Butt 2013a; Buxton et al. 2011; Houston 2005). However, an increasing bifurcation is apparent between the large-scale operations in industries such as intensive poultry farms and the many small-scale, sub-commercial farms in industries such as cattle farming (Butt 2013b).

These changes have occurred in a demographic environment in which Australia's largest cities continue to grow in terms of population and physical size, though trends toward counterurbanization in some rural areas exist alongside depopulation of many inland agricultural regions. Australia's cities exhibit very low population densities by international comparison but are similar to those of many North American cities. Over 80 percent of Australians live in urban areas, typically in low-density, single, detached houses in suburban settings—a long-standing Australian urban pattern (Davison 2006). Despite policy attempts to encourage higher-density housing in urban areas, Australia's cities continue to sprawl into surrounding, previously rural environments and frequently onto areas of productive farmland. In addition, the spread of non-agricultural land uses into peri-urban and coastal amenity regions has created a diffusion of competing land markets and activities, diluting and removing agricultural activities in many of these landscapes. This contest between land uses is at the heart of much of the land-use planning agenda for peri-urban and rural areas in Australia. Several commentators provide perspective on this conflict between practice and policy:

Development at the rural urban interface can be considered from various perspectives, but the debate is generally dominated by issues associated with urbanisation, such as the lack of infrastructure. The importance of peri-urban agriculture for a healthy city and the advantages of retaining agriculture are being increasingly recognised in rhetoric, but not in planning strategies, although the urbanisation of agriculturally productive land is of concern around many cities. (Parker and Jarecki 2003)

Throughout Australia, while broad objectives for the future of peri-urban areas as rural (and agricultural) places are articulated at a planning policy level, land use realities, along with market preferences and public and private investment (particularly in transport), are often in conflict with the effective implementation of planning objectives. Processes of rural land use dilution and agricultural displacement continue to occur, resulting in larger populations living beyond the urban fringe, with an increasing concentration of this population in areas that are accessible to metropolitan and regional cities. (Butt 2013b, 426)

During the last decade the respective State governments in Australia have prepared new or revised comprehensive metropolitan planning strategies. These strategies have increasingly had at their core policies and initiatives to control and limit urban sprawl. These objectives have generally been driven by concerns over the implications of such a pattern for transport access and costs and the provision of infrastructure, facilities, and services, rather than a deep concern about encroaching on productive agricultural land. There has been some recognition in Australia's metropolitan strategies of the significance of agricultural production at the urban edge and the need for urban form and transport planning to consider access to food supplies but none of them have given more than modest consideration to this issue. (Budge 2013, 370)

Australian agriculture, as with much of the population settlement pattern, is largely predicated on the nature of the landscapes and sustainable water supplies in what is the world's driest and oldest continent. Lesslie, Mewett, and Walcott describes the Australian context as follows:

> The total area of land under agricultural land uses in Australia in 2006 was 4.57 million square kilometres comprising over 59 per cent of the continent. . . . The dominant land use overall was livestock grazing on natural vegetation in arid and semi-arid regions (3.56 million square kilometres or 46 per cent). Grazing on modified pastures made up 9.5 per cent (or 730 000 square kilometres) of land use. A much smaller proportion of the land area was taken up by other agricultural uses, including broadacre[1] cropping (270 000 square kilometres or 3.5 per cent) and horticulture (5000 square kilometres or 0.06 per cent). (Lesslie, Mewett, and Walcott 2011, 3)

Australia's major cities and towns are almost all located along the coastal fringes, with over 80 percent of all Australians living within fifty kilometres of the coast (Australian Bureau of Statistics 2012a). The vast majority of the continent receives irregular seasonal rainfall, with an annual average well below 400 millimetres. Conversely, parts of temperate southeast Australia typically experience higher rainfall and annual snowmelt in some river catchments, while in some areas of northern Australia annual average rainfall in excess of 2,000 millimetres is typical within tropical weather systems (Bureau of Meteorology 2008).

The issue faced by much of Australia is the seasonal unreliability of rainfall, with a consequent reliance in many areas on irrigation water for farming of crops or pastures. Additionally, in many areas, there is now an overreliance on irrigation water, a reliance compounded in times of extended drought.

NATIONAL AND REGIONAL POLICY FRAGMENTATION

As a continental land mass, Australia has the advantage of a single national government to manage the resource base, a unique situation in the world. Yet, regarding many areas of natural resource stewardship, Australia has not really grasped the many opportunities that this situation provides. The Australian Constitution leaves land-use planning, environmental management, and urban development, which so often have major impacts on productive farmland, as functions performed by the respective state governments. The extended period of debate among the six former British colonies that resulted in formation of the Commonwealth of Australia in 1901 meant that powers of the national government were limited to the roles and responsibilities that the former colonies (now states) were willing to secede to the new national government. As a result, the commonwealth government has almost no jurisdiction over or legal responsibility for the regulation of land use and environmental management or agricultural policy outside trade and export. Amendment of the Australian Constitution is a difficult and protracted process, and most proposed changes have failed to achieve the necessary popular support through referendum. The current arrangements and division of responsibilities for land-use policy are therefore likely to frame the foreseeable future, and there is little likelihood that there will be any changes to this division of powers.

The only real means that the commonwealth government has for effectively exerting an influence on large-scale land-use planning, allocation, or regulation of resources, or environmental management, is through tied financial grants, monetary incentives, or subsidies to the states, regional bodies, local governments, or community-based groups. Traditionally, most commonwealth governments have shied away from any direct involvement in land-use planning, including the management of metropolitan growth. Consequently, there has been no national framework for how policy and regulative issues should apply and whether productive farmland should be protected from conversion to other uses as a national priority (Gardner 1994; Houston 2005; Sinclair 2002).

Historically, settlement and agriculture have developed generally around the large Australian coastal cities. From the commencement of European urbanization in the nineteenth century, this was a consequence not only of the need to have access to urban markets but also of urban settlements in areas of higher, more regular rainfall and better soils for domestic and export production. The consequences of this pattern and approach are starting to be recognized nationally; the Sustainable Australia Report found that "the rapid outward growth of our cities presents significant environmental, social, and economic issues. Historically, major settlements (now some of our largest cities) were located in fertile areas with mild climates and abundant natural resources. Food production close to the settlement was important to feed growing townships. As our major cities have grown, they have displaced agricultural activities and built over productive farming land" (National Sustainability Council 2013, 58). Agricultural expansion from the mid-nineteenth century included the development of access to inland, dryland grazing areas, cereal cropping in areas of marginal rainfall, the eventual development of irrigation farming in parts of the Murray Darling Basin (Australia's largest river basin) in the southeast, and the emergence of tropical and subtropical farming in the coastal north (Roberts 1968). Much of this expansion was the result of a strongly state-directed approach to both agriculture and regional development. Examples such as "closer settlement" schemes (often linked to soldier settlement programs), land grants, the development of rail infrastructure, and localized agricultural extension and training in various forms all typified policy approaches from the mid-nineteenth century to the mid-twentieth century. In many instances, the limits to productivity in a low-rainfall environment were quickly uncovered, often with significant economic and personal costs (Barr and Cary 1992).

Access to water resources has become central to the geography of much of the agriculture in Australia. However, areas with better rainfall (aside from tropical Australia) typically coincide with areas of urban settlement and peri-urban expansion. Conversely, vast areas of inland Australia rely on the large scales of farms and farm businesses, as well as evolving agronomic practices (Barr and Cary 1992), to overcome inherent limits to productive capacity and yields in low-rainfall cropping and grazing environments. Irrigation, notably but not exclusively in the Murray Darling system, offers a third alternative, but water volume and reliability have remained problematic despite continued technological and regulatory improvements (Musgrave 2011). A long-running national, state, and regional debate about competing water interests culminated most recently in the politically and communally contentious Murray Darling Basin Plan (Murray Darling Basin Authority 2012). The plan aims to allocate limited water resources among competing agricultural, environmental, and communal uses across parts of three Australian states where water allocation is the difference between highly productive farmland and dryland pastures or crops left to the vagaries of an unreliable and low-rainfall regime.

Industry support and farm business assistance are regular features of public policy at the national level. The commonwealth government's involvement in land management issues has increasingly been in response to broad political and communal concern and direct representation by the states and regions and has usually involved financial packages aimed at major environmental issues. This has involved large-scale resource management and landscape change such as securing long-term water supplies in the world's driest continent, now facing the prospect of ongoing lower rainfall because of climate change, supporting vegetation and habitat management, and addressing the increasing areas of land subject to salinity. These are both agricultural industry concerns and a more "post-productivist" turn in policy in which environmental values and rationales now figure more strongly in policy debates than was the case in previous decades.

The emerging national and international agendas on food security, in its many aspects, led the Australian government to promise during the 2010 national election to develop Australia's first "National Food Plan." As a consequence, an issues paper was released in mid-2012, and a consultation program was organized with the intention that a final National Food Plan would assist to "foster a sustainable, globally competitive, resilient food supply that supports access to nutritious and affordable food" (Department of Agriculture, Fisheries, and Forestry 2012, 13–14). Included in this proposed plan is the objective to "maintain and improve the natural resource base underpinning food production in Australia" (ibid.). This is as close as the background documents have come to a specific reference to the concept or value of preserving productive farmland from urban encroachment or competing resource demands. At best, it is noted, "the effect of urban displacement of food production is not well understood and is too early to assess the effect if any on the food sector or consumers" (ibid.).

Despite this lack of focus on farmland preservation in the government's presentation of the issues during the consultation period, the preservation of productive farmland was consistently raised in both written submissions and at public forums. In the context of food security, there was an identified concern in the feedback about both "increasing competition for arable farm land from urban encroachment, such as vegetable farms close to cities being taken over by housing" (this related particularly to horticultural land at the urban fringe), and "increased competition for land use from mining exploration and production in some areas" (Department of Agriculture, Fisheries, and Forestry 2012, 64).

In its submission to the issues paper, the Planning Institute of Australia (the national professional association of urban and regional planners) noted that "the land that is needed to grow [this] food is being eroded by a number of pressures for change in the land use to other types of development. This includes urbanisation and rural residential development in the metropolitan fringe and coastal areas, as well as mining and gas extraction in the inland areas, particularly amongst the important cropping lands." The submission advocated the "need to strategically identify lands for cropping in particular in the inland, and the lands for growing perishable vegetables on the Metropolitan fringe and coastal areas. These lands need to be protected from the urbanisation and rural residential development on one hand and resource extraction on the other" (Planning Institute of Australia 2012).

Such views have not been confined to pressure or lobby groups. A recent Senate inquiry into food production in Australia made the following observations:

> Land available for agriculture is declining across the globe as expanding populations inhabit fertile land that could otherwise be devoted to food production. Although this problem is not as severe in Australia as it is in countries with a smaller land mass, urban encroachment is nonetheless affecting the capacity of Australian producers to grow food in the areas in which it is demanded, which in turn affects its quality and affordability. Competition for fertile land from mining and biofuels also threatens to reduce Australia's productive capacity. Australian governments need to give serious consideration to mechanisms for protecting our most fertile agricultural land from alternative uses in the interests of our long-term productive capacity and food security. (Senate Select Committee 2010, 20–21)

RECENT INITIATIVES BY STATE GOVERNMENTS TO PROTECT FARMLAND

It is against this background of the national government's reluctance to address the issue of farmland preservation that Australia's initiatives on farmland protection and preservation need to be studied and assessed. Land-use planning, in both a policy and a regulatory sense, is essentially a responsibility of the six state governments. Their approach to farmland preservation has been understandably somewhat disjointed and lacked a consistent theme. In turn, the six state governments have largely delegated the day-to-day administration of planning schemes and development control plans to local governments. The hope in some quarters that a National Food Plan would address this issue nationally now appears unlikely to be fulfilled.

Figure 13.1. Priority issues from the National Food Plan Green Paper 2012 public meetings.

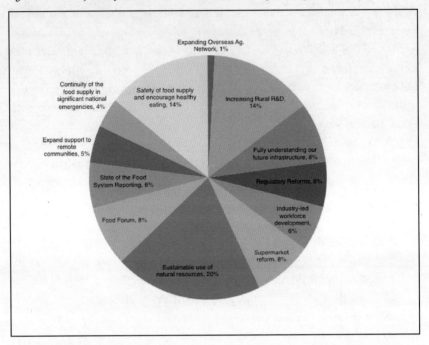

Note: The top three priority issues identified from the National Food Plan Green Paper consultations nationally were: (1) ensuring Australia's natural resources, especially land and water, are used sustainably (20 percent); (2) increasing rural research and development investment over the next ten years (14 percent); (3) further improvements to the safety of our food supply and encourage healthy eating (14 percent). See http://www.agriculture.gov.au/ag-farm-food/food/publications/national_food_plan/green-paper.

The other circumstance that contributes to understanding Australia's disjointed and reactive approach to farmland preservation is the relatively localized nature of population pressures on the land. In a national land-use context, there are relatively small productive areas adjoining the five largest metropolitan areas. However, with much of this farmland focused on horticulture, the value of such production is significant nationally, with some estimates (Houston 2005) suggesting that it is worth more than 20 percent of national agricultural production. An increasing proportion, though, is "shed based," relying on proximity to labour forces and transportation to the metropolitan areas rather than the inherent soil quality and growing conditions.

Although historically there has been limited coordinated action at national and state levels on farmland preservation, it is apparent that over the past two decades or so there has been a significant increase in attention to farmland protection, particularly at the edges of the major metropolitan areas, with varying degrees of commitment and success. Measures have ranged from fairly vague state policies with inconsistent implementation to attempts at legislative requirements. However, neither has been predicated solely on farmland preservation.

Agitation to preserve farmland has essentially derived from increasing concerns by a coalition of land-use planners, agricultural resource practitioners, farmers' groups, and conservationists' lobbying that a limited resource is under threat and that the loss of highly productive agricultural land will have profound impacts on future levels and costs of food production and transportation and therefore on food prices (Budge 2013; Carey et al. 2011; Houston 2005). Despite Australia's relatively large land mass and apparent capacity to bring new land under production, the limits to intensive agriculture are being reached, particularly where it is reliant on irrigation. The permanent removal of productive land is increasingly being seen as contrary to Smart Growth planning for cities and towns (Budge 2013). However, one of the enduring legacies of a "pioneer" nation such as Australia is widespread and continuing belief that there will always be more land available and that technology will continue to deliver in the pursuit of greater levels of production. The need to take tough regulatory stances to preserve and manage land has been a difficult message to convey to the wider community and to those charged with long-term land-use planning and resource management.

Generally, state government efforts at policy formulation on retaining productive farmland can be characterized as largely aspirational. These policies have usually lacked real implementation and, when confronted by strong forces of urban development, often been put aside. Specific regulatory measures on the ground have often been watered down or lost under the pressure of landowners who can see the march of new housing estates coming over the hill and farmers looking for windfall retirement savings boosts by cashing in their landholdings (Buxton and Goodman 2003). Johnson, Kelleher, and Chant discuss the challenges facing agriculture in the Sydney region:

> The physical environment and growth of Sydney as a major centre have combined to create a dramatic example of the pressures on fringe agriculture. The Sydney region is effectively locked into a basin by geographical features, the Blue Mountains to the west, river and catchment systems to the north and south, and the Pacific Ocean to the east. Agriculture

has been forced to compete for a finite supply of space with other land uses, most demonstrably that of urban development and associated planning, but also increased community pressures for landscapes dedicated to natural environmental purposes, and free from agricultural chemicals. (Johnson, Kelleher, and Chant 1998, n.p.)

Increasingly and largely because of the growing complexity of managing metropolitan growth, state governments find it necessary to be more interventionist in terms of key elements that shape and direct the metropolitan urban form. State governments are establishing longer-term goals and requirements for infrastructure and the constraint on or containment of urban sprawl. The primary motivation in initiating and framing metropolitan strategies is the requirement to address the infrastructure and service needs of a growing population and particularly the land consumption implications of projected increases in household numbers. The issue of productive farmland has emerged as only *one* of the factors influencing metropolitan strategies (Budge 2013).

State governments have a number of means by which they can set directions for land-use policy. Mostly, they derive from a capacity within their legislative powers to specify a state land-use policy. Most states have taken up this opportunity in the past two decades to define a policy on retaining productive agricultural land. Set out in Table 13.2 is a summary of the initiatives that each state has pursued on farmland protection.

Although each state has a policy framework for decision making, there is no consistency among the states. The level of detail, the factors considered, and the level of compliance with policy vary. In all but South Australia, state governments have adopted a statewide policy or set of provisions designed to put some common requirements into the consideration of whether land should be rezoned or whether changes in land use or new developments should be supported that would permanently remove land from agricultural production. Generally, these policies seek to recognize agricultural land as a significant commodity and valued economic resource. These policies also set out a number of criteria that should be met but are not mandatory. Generally, they are applicable only when there are development proposals or rezonings that would permanently remove productive or high-quality agricultural land from agricultural use (Budge et al. 2012).

Table 13.2. Initiatives pursued by states on farmland protection.

STATE	(1) RESPONSE TO AGRICULTURAL LAND PROTECTION (2) POLICY INITIATIVES (3) LEGISLATION
NEW SOUTH WALES	(1) STATE ENVIRONMENTAL PROTECTION POLICY (2) DEPARTMENT OF PRIMARY INDUSTRIES POLICY FOR PROTECTION OF AGRICULTURAL LAND
QUEENSLAND	(1) STATE PLANNING POLICY (2) CONSERVATION AND DEVELOPMENT OF AGRICULTURAL LAND (3) STRATEGIC CROPPING AREAS LEGISLATION
SOUTH AUSTRALIA	(1) NO SPECIFIC STATE POLICY (2) STATE STRATEGY REFERENCES TO AGRICULTURAL LAND (3) DESIGNATION OF TWO CHARACTER PRESERVATION AREAS
TASMANIA	(1) STATE POLICY ON HIGH-QUALITY AGRICULTURAL LAND (2) POLICY REQUIREMENTS INCORPORATED IN ALL PLANNING SCHEMES (3) RIGHT-TO-FARM LEGISLATION
VICTORIA	(1) STATE PLANNING POLICY (2) POLICY REQUIREMENTS INCORPORATED IN ALL PLANNING SCHEMES (3) URBAN GROWTH BOUNDARY LEGISLATION
WESTERN AUSTRALIA	(2) STATEMENT OF PLANNING POLICY

Source: Author's Summary.

There are three broad distinguishing features of these state policies. Each attempts to provide a different definition of what constitutes productive or high-quality land. Each provides for different levels or scales of factors to be considered, and each is subject to different levels of inputs to the decision-making process by different bodies, such as government departments and agencies with an interest in the outcome (Budge et al. 2012).

Two significant recent state initiatives have been the Queensland government's Strategic Cropping Land Act 2013 and the South Australian government's Character Preservation Acts that apply to two designated areas. On the surface, these seem to be major initiatives in farmland preservation and protection. However, an analysis indicates that they might have limited capacities to preserve farmland.

The objectives of the Strategic Cropping Land Act focus on protecting land highly suitable for cropping, managing the impacts of development on that land, and preserving the productive capacity of that land for future generations. The act seeks to achieve these objectives by

- identifying potential strategic cropping land (SCL);
- providing criteria to decide whether or not land is SCL;
- establishing protection and management areas;

- providing for development assessment;
- imposing conditions on development;
- preventing permanent impacts on SCL in protection areas (unless the development is in exceptional circumstances); and
- requiring mitigation to be paid by developers if SCL is permanently impacted in the management area or by a development in exceptional circumstances.

The *State Planning Policy 1/12: Protection of Queensland's Strategic Cropping Land* (Queensland Government 2011, 20–21), which accompanies the act, identifies the need to protect strategic cropping land because "it is a finite resource that must be conserved and managed for the longer term. As a general aim, planning and approval powers should be used to protect SCL from those developments that lead to permanent impacts or diminished productivity." However, the explanatory material accompanying the act and policy makes it clear that designation of areas as "strategic cropping" does not "lock up" these areas; rather, such a designation ensures that their role in and value for agriculture must be seriously addressed in any proposal for conversion to a non-agricultural use (Department of Environment and Resource Management 2011). Already fear is expressed in some farming communities that the short-term financial gains from mining will be pre-eminent in conflicts over land use.

In South Australia, the Department of Planning and Local Government has expressed the need "to protect the special character of the Barossa Valley and McLaren Vale" by pursuing a range of objectives:

- To recognize, protect, and enhance the special character of the district while at the same time providing for the economic, social, and physical well-being of the community.
- To ensure that activities that are unacceptable in view of their adverse effect on the special character of the district are prevented from proceeding.

The character values of the two districts are recognized as

a. the rural and natural landscape and visual amenity of the district

b. the heritage attributes of the district

c. the built form of the townships as they relate to the district
   d. the viticultural, agricultural, and associated industries of the district
   e. the scenic and tourism attributes of the district.
   (Department of Planning and Local Government 2011)

Although the legislation has yet to be tested, it is evident from the explanatory material and the government publicity that the value of tourism rather than the value of farmland is the underlying goal of preservation efforts there. The iconic values of these two areas, their contributions to the amenity of the expanding Adelaide area, and the economic values of these areas for tourism are the prime drivers rather than a focus on preserving farmland because of its inherent value and quality.

CASE STUDY: THE MELBOURNE METROPOLITAN STRATEGY AND FARMLAND PRESERVATION

In Victoria, a new state government was elected in late 2010 partly on the platform of preparing a new comprehensive land-use strategy for the Melbourne metropolitan area. Similar to the consultation on a metropolitan strategy in the early 2000s, one of the major issues raised by the community was the need to curtail urban sprawl and retain the natural resource and landscape setting of Melbourne. That setting includes the urban area adjoining some areas used for highly productive horticultural farming. Many areas surrounding the Melbourne metropolitan area remain highly productive agricultural regions, with output per hectare up to four times the state average (Carey et al. 2011). Industries such as horticulture are concentrated in areas of high-quality soils, while activities such as intensive poultry raising are reliant on transportation, processing, and labour force proximity. However, the ongoing viability of these industries is under pressure not only from urban expansion but also from rural dilution, with a growing prevalence of small and sub-commercial (hobby farming) activities in the broader region (Butt 2013b). A lack of integration between urban development and agricultural land management policy has been apparent for decades. Few exceptions exist, and they typically relate to areas of identified high-amenity and landscape values (Budge et al. 2012).

Melbourne 2030: Planning for Sustainable Growth (Department of Infrastructure 2002) was prepared primarily to provide a clear direction to and prescription for how an anticipated 620,000 new households to the year 2030 could be accommodated. By using a combination of mixed-use activity centres, retrofitting existing suburbs, and restoring "brownfield" sites, housing at the urban fringe was to be limited to 31 percent of the total projected numbers. Urban growth at the metropolitan fringe was virtually confined to three major corridors, thereby mostly retaining the extensive greenbelt that largely encircles Melbourne. The Melbourne 2030 strategy built upon a succession of metropolitan strategies developed for the area since the late 1960s. Buxton and Goodman (2002) trace the historical development of Melbourne's greenbelt from its original establishment as part of the overall metropolitan strategy at the time and as a keystone in the Melbourne metropolitan strategy within the evolution of international thinking on development of the concept of greenbelts or green wedges, as they were termed. The non-urban policies and zones of the 1970s encouraged agriculture as well as a range of other non-urban activities, such as nature conservation and extractive industry: "However, in the last twenty-five years agriculture has changed around the metropolitan area. Many orchards and intensive horticulture areas have moved to other locations, dairying has virtually disappeared and part-time farmers, as opposed to persons earning a full time living from the land, now occupy broad acre farming areas" (Alastair Kellock and Associates 2000, 41).

A succession of state government–prepared metropolitan strategies during the 1970s and 1980s confirmed the green wedge framework until the mid-1990s. That strong principle was eroded by a series of decisions by a state government committed to a market force approach to planning. Buxton and Goodman (2002) identified that some 4,000 hectares of land in the green wedges were rezoned for residential development between 1996 and 2002.

It is important to note when discussing development of the succession of Melbourne metropolitan strategies, and particularly the Melbourne 2030 strategy, that they were not designed as comprehensive initiatives to address the protection of productive farmland at the metropolitan edge and the immediate hinterland. Land set aside in the green wedges is not solely devoted to agriculture. The Melbourne 2030 strategy recognizes the multi-purpose nature of these areas. Land within the green wedges is set aside for designated uses such as the Melbourne International Airport, sewage treatment works, quarries, and extensive wetlands and nature conservation areas.

When the Melbourne 2030 strategy was announced, there was considerable speculation about how the proposed urban growth boundary, together with the twelve designated green wedge areas, would or even could be strictly enforced. The Victorian planning system provides that all amendments to planning schemes be approved by the minister for planning. But even that "protection," as Buxton and Goodman (2002) reported, had seen land in the green wedges progressively consumed for new residential developments. In a bold move, and to the consternation of much of the land development and speculation industry, the government announced that it would pass legislation defining the urban growth boundary. Amendments to the Victorian state planning legislation, the Planning and Environment Act 1987, in February 2003 introduced a new part to the act that defined a planning scheme for the metropolitan fringe, an urban growth boundary, and green wedge land (land outside an urban growth boundary). All changes to the provisions by the minister for planning *must be ratified by Parliament.*

The urban growth boundary imposed on each of the seventeen local planning schemes that cover the urban fringe has also been placed around thirty-six small townships that lie beyond the contiguous metropolitan area but inside the designated Melbourne metropolitan planning area. To further reinforce protection of the green wedge areas from inappropriate or intrusive development, the government used the powers available to it to impose a series of standard zones on all lands in the green wedges. Even without a specific stated intention to preserve farmland, implementation of the Melbourne strategy has potentially delivered the most significant agricultural protection measure anywhere in Australia. Effectively, through legislative measures, all lands, whether productive or not, outside the declared growth corridors cannot be converted to urban uses without an act of Parliament. Significantly, the defined metropolitan area of Melbourne for planning purposes has always extended well beyond the actual built-up area. The contiguous urban area encompasses less than 50 percent of the total defined metropolitan area. The means of enforcement through a legislated urban growth boundary are generally considered to be some of the strictest provisions in metropolitan planning around the world (Buxton and Goodman 2003). The Victorian minister for planning has effectively removed the position's discretionary powers.

In May 2004, to further reinforce the purpose and role of the green wedges, the state government imposed new zones across those parts of the seventeen municipalities with lands in the designated green wedges. The specific green wedge zone recognizes the diversity of land uses in the green wedge areas but also clearly recognizes that agriculture is an important use to be protected. These actions in relation to Melbourne were essentially driven by state government policy initiatives designed to limit urban sprawl, not to protect farmland. These measures were largely in response to demands for services and infrastructure and a strong community-driven campaign to protect Melbourne's green wedges.

Despite the stated intentions and barriers that the government put in the way of enlarging the metropolitan area at the expense of these green wedges, which included preserved farmland, a succession of state governments acquiesced to developers' interests in the face of unprecedented population growth, increasing housing prices, and the call to release more land for housing. These changes to metropolitan strategy in this instance have weakened the initial intent of this policy while responding to issues of urban housing affordability and urban land supply in an already compromised peri-urban land market. Carey et al. (2011) note that the Melbourne region accounts for 72 percent of Victoria's vegetable production, yet attempts to control the urban expansion of Melbourne have been largely ineffective despite a legislated urban growth boundary. In 2010, the Victorian legislature approved changes to the boundary by adding 43,600 hectares to provide twenty years of additional land supply for new housing. This was the third time that the boundary had been moved since its introduction in 2002, even though, when it was originally legislated, it was on the basis that it already provided sufficient land for twenty years. Carey et al. (2011) commented that this boundary extension would significantly diminish the amount of "prime food-growing land" at the city's fringe and that soaring land values in these peri-urban areas would limit farm expansion and investment. In 2012, the state government formally announced its intention to undertake major revisions to the metropolitan planning strategy and issued a discussion paper in which, expectedly, mention of farmland preservation was limited. The only reference to this issue stated that "some areas around Melbourne contain highly productive agricultural land. The Port Phillip and Western Port regions are the second highest producers of agricultural products in Victoria, with agricultural output per hectare approximately four times the state average. Agriculture around Melbourne is a declining proportion of local economic activity and employment, however the quantity of agricultural production remains significant. Many types of agriculture occurring in peri-urban areas (such as aquaculture, poultry and egg farming and some types of horticulture) can be undertaken in areas without high quality soils" (Department of Planning and

Community Development 2012). It is unlikely that farmland preservation will be a major driver of the new metropolitan strategy.

CONCLUSION

The protection and preservation of productive farmland in Australia has had a long gestation period. Largely ignored as an issue of national importance and lacking a framework to provide a consistent approach among the six state jurisdictions, the voice of farmland preservation is increasingly being heard in a range of forums and at various levels, yet with no apparent overarching narrative or coherent framework. The issue has been raised to some significance but largely in relation to wider discourses of food security (and "sovereignty"), the environmental impacts of mining, and the amenity values of peri-urban landscapes, rather than in relation to productive concerns. In Australia, land competition is often concentrated in the most productive and populated areas, despite a low population and an expansive land area. Houston's (2005, 19) comment that there is "a national tug-of-war, albeit undeclared, over the allocation of key natural resources" remains relevant today. Houston adds that, "because peri-urban regions will be the area of significant population growth for the foreseeable future, these competing interests need to be more actively and deliberately mediated. Amongst other things, successfully mediating the interface between urbanisation and agriculture will become increasingly important."

The protection of land used for agriculture in Australia, especially highly productive land, is a story of tentative measures by each of the state governments with varying degrees of effectiveness. The protection implemented has occurred primarily as part of state-initiated metropolitan strategies developed in attempts to limit urban sprawl. Those strategies have chosen land for future development with little regard for its productive capacity. It is apparent, though, that the argument for farmland preservation has been considerably strengthened where a range of social, cultural, lifestyle, and recreational arguments has been introduced. Importantly, protecting productive potential does not appear to have been effective as a goal in itself; the multifunctional role of such land in reinforcing the aesthetics of the metropolitan area remains significant. To this long-standing issue has been added the more recent clashes among farming, mining, and water for agriculture and the environment: "Land use change and land management [are] central to current debate in Australia around food security, water, climate change adaptation, population, and urban expansion" (Lesslie and Mewett 2013, 1).

The Australian land-use planning system is generally structured in such a way that the states have sufficient powers to implement and enforce strong measures for protecting and managing productive agricultural lands. However, their performance has been inconsistent. As well, successive national governments simply have not seen this objective as a priority. Time will tell if innovative and bold steps such as a National Food Plan or Queensland's Strategic Cropping Land Act are forerunners to new directions in Australian land-use management and farmland preservation. The indications to date are that Australia is still well short of a comprehensive and systematic process to tackle this increasingly important issue.

NOTES

1. "Broadacre" is a term used, mainly in Australia, to describe farms or industries engaged in the production of grains, oilseeds, and other crops (especially wheat, barley, peas, sorghum, maize, hemp, safflower, and sunflower), or the grazing of livestock for meat or wool, on a large scale using extensive parcels of land.

REFERENCES

ABARES (Australian Bureau of Agricultural and Resource Economics and Sciences). 2007. *Drivers of Change in Australian Agriculture*. Publication No. 07/057. Canberra: RIRDC.

———. 2012. *The Agricultural Commodities Statistics Report*. Canberra: ABARES. http://www.daff.gov.au/abares/data.

Alastair Kellock and Associates. 2000. *Green Wedges and Non-Urban Issues Technical Report 2*. Melbourne: Department of Infrastructure.

Australian Bureau of Statistics. 2010. *Agricultural Commodities Australia 2008–09*. Canberra: Australian Bureau of Statistics.

———. 2012a. *Census of Population and Housing—2011*. Canberra: Australian Bureau of Statistics.

———. 2012b. *Gross Value of Irrigated Agricultural Production, 2010–11*. Canberra: Australian Bureau of Statistics.

Barr, N., and J. Cary. 1992. *Greening a Brown Land: The Australian Search for Sustainable Land Use*. Melbourne: Macmillan Education.

Budge, T. 2013. "Is Food a Missing Ingredient in Australia's Metropolitan Planning Strategies?" In *Food Security in Australia: Challenges and Prospects for the Future*, edited by Q. Farmar-Bowers, V. Higgins, and J. Miller, 367–79. New York: Springer.

Budge, T., A. Butt, M. Chesterfield, M. Kennedy, M. Buxton, and D. Tremain. 2012. *Does Australia Need a National Policy to Preserve Agriculture Land?* Sydney: Australian Farm Institute.

Bureau of Meteorology. 2008. *Climate of Australia*. Melbourne: Bureau of Meteorology.

Butt, A. 2013a. "Exploring Peri-Urbanisation and Agricultural Systems in the Melbourne Region." *Geographical Research* 51 (2): 204–218.

———. 2013b. "Development, Dilution, and Functional Change in the Peri-Urban Landscape: What Does It Really Mean for Agriculture?" In *Food Security in Australia: Challenges and Prospects for the Future*, edited by Q. Farmar-Bowers, V. Higgins, and J. Miller, 425–41. New York: Springer.

Buxton, M., and R. Goodman. 2002. *Maintaining Melbourne's Green Wedge: Planning Policy and the Future of Melbourne's Green Belt*. Melbourne: School of Social Science and Planning, RMIT University.

———. 2003. "Protecting Melbourne's Green Belt." *Urban Policy and Research* 21 (2): 205–10.

Buxton, M., A. Butt, S. Farrell, and A. Alvarez. 2011. "Future of the Fringe: Scenarios for Melbourne's Peri-Urban Growth." Melbourne: State of Australian Cities Conference.

Carey, R., F. Krumholz, K. Duignan, K. McConell, J.L. Browne, C. Burns, and M. Lawrence. 2011. "Integrating Agriculture and Food Policy to Achieve Sustainable Peri-Urban Fruit and Vegetable Production in Victoria, Australia." *Journal of Agriculture, Food Systems, and Community Development* 1 (3): 181–95.

Davison, A. 2006. "Stuck in a Cul-de-Sac? Suburban History and Urban Sustainability in Australia." *Urban Policy and Research* 24 (2): 201–16.

Department of Agriculture, Fisheries, and Forestry. 2012. *National Food Plan Green Paper 2012*. Canberra: Department of Agriculture, Fisheries, and Forestry.

Department of Environment and Resource Management. 2011. *Protecting Queensland's Strategic Cropping Land: Draft State Planning Policy*. Brisbane: Department of Environment and Resource Management. http://www.derm.qld.gov.au/land/planning/pdf/strategic-cropping/scl-draft-spp.pdf.

Department of Infrastructure. 2002. *Melbourne 2030: Planning for Sustainable Growth*. Melbourne: Department of Infrastructure.

Department of Planning and Community Development. 2012. "Let's Talk about the Future: A Discussion Paper Prepared by the Ministerial Advisory Committee for the Metropolitan Planning Strategy for Melbourne." Melbourne: Department of Planning and Community Development. http://www.planmelbourne.vic.gov.au/discussion-paper.

Department of Planning and Local Government. 2011. "Protecting the Barossa Valley and McLaren Vale." Discussion paper. Adelaide: Government of South Australia.

Gardner, B. 1994. "Highly Productive Agricultural Land: Australia's Limited Resource." In *Agriculture and Rural Industries on the Fringe*. Bendigo: Australian Rural and Regional Planning Network and TBA Planners.

Houston, P. 2005. "Revaluing the Fringe: Some Findings on the Value of Agricultural Production in Australia's Peri-Urban Regions." *Geographical Research* 43 (2): 209–23.

Johnson, N., F. Kelleher, and J. Chant. 1998. "The Future of Agriculture in the Peri-Urban Fringe of Sydney." Hawkesbury: Farming Systems Research Centre, University of Western Sydney. http://www.regional.org.au/au/asa/1998/7/069johnson.htm.

Lesslie, R., and J. Mewett. 2013. "Land Use and Management: The Australian Context." Research Report 13.1. Canberra: Australian Bureau of Agricultural and Resource Economics and Sciences.

Lesslie, R., J. Mewett, and J. Walcott. 2011. *Landscapes in Transition: Tracking Land Use Change in Australia*. Science and Economic Insights Issue 2.2. Canberra, Australia: Australian Government Department of Agriculture, Fisheries and Forestry.

Mullen, J. 2010. "Agricultural Productivity Growth in Australia and New Zealand." In *The Shifting Patterns of Agricultural Production and Productivity Worldwide*, edited by J. Alston, B. Babcock, and P. Pardey, 99–122. Ames: Iowa State University.

Murray Darling Basin Authority. 2012. *Murray Darling Basin Plan*. Canberra: Government of Australia.

Musgrave, W. 2011. "Historical Development of Water Resources in Australia: Irrigation Policy in the Murray Darling Basin." In *Water Policy in Australia*, edited by L. Crase, 28–43. Washington, DC: Earthscan.

National Sustainability Council. 2013. *Sustainable Australia Report 2013: Conversations with the Future*. Canberra: Government of Australia.

Parker, F., and S. Jarecki. 2003. "Transitions at the Rural/Urban Interface: 'Moving in,' 'Moving out,' and 'Staying Put.'" In *Proceedings of the State of Australian Cities Conference*. Parramatta.

Planning Institute of Australia. 2012. Submission to the *National Food Plan Green Paper 2012*. Canberra: Climate and Health Alliance.

Prime Minister's Science, Engineering, and Innovation Council. 2010. *Australia and Food Security in a Changing World*. Canberra: Prime Minister's Science, Engineering, and Innovation Council.

Queensland Government. 2011. Strategic Cropping Land Act. http://www.legislation.qld.gov.au/LEGISLTN/ACTS/2011/11AC047.pdf.

Roberts, S. 1968. *History of Australian Land Settlement 1788-1920*. 1924; reprinted, Melbourne: Macmillan.

Senate Select Committee. 2010.

Sinclair, I. 2002. "Preserving Rural Land in Australia." Paper presented to the Joint Royal Australian Planning Institute/New Zealand Planning Institute Conference, Wellington.

CONCLUSION

Farmland Preservation: Land for Future Generations

WAYNE CALDWELL, BRONWYNNE WILTON, AND KATE PROCTER

A generation from now there will be a postmortem on how well the current generation has done in protecting the important land resources that support agriculture. Although agriculture exists in many forms and within different contexts, quality soils, which exist in favourable climates, will always be of the highest priority for protection. Unfortunately, these lands often exist close to growing urban centres and are equally desired for high-value urban uses.

There is also an impending calamity that elevates the urgency of this issue. Climate change models suggest that, while across the globe there will be areas of increased productivity, there will be even greater areas that suffer long-term losses in productivity. This raises many difficult questions. What does it mean from a moral and ethical perspective to reside in an area of increased productivity while much of the rest of the planet faces daunting challenges ranging from a diminished agricultural sector to literal starvation? What roles do governments and individuals play? What are their responsibilities? Which tools of farmland preservation are most appropriate for different contexts and geographies? And, ultimately, how well have we done in protecting farmland for future generations?

This book has provided many perspectives, tools, and strategies to help with the protection of farmland.

As civilizations have struggled throughout the centuries to survive and thrive, the land needed to sustain life remains most valuable. Preserving and protecting this resource vital to producing food, one of the most basic necessities of life, remain keys to national security, even if most of the population does not recognize this. In part because large-scale scarcity in North America has not come to pass as predicted, it has sometimes proven difficult to argue the case for farmland preservation.

However, competing demands for the land continue to increase as populations grow and become ever more complex. Land-use competition demands that we pay attention and protect this most valuable of resources so that it remains productive and able to sustain life. Although losses of farmland up to this point have been more than compensated for by increases in cropping efficiency and productivity, reliance on unlimited growth in agricultural production for future needs might not prove to be in our collective best interest.

Chapter 1 argued that the loss of farmers rather than the loss of land should be of utmost concern. As farmland has become increasingly consolidated since 1941, there has been a reduction of 71 percent in the number of farms. During the same time, the farmland base has decreased by only 8 percent, and the area used for crops has actually increased by 56 percent. Thus, farmland preservation on a broad scale might be more dependent on recreating economically viable family farms.

Considering farmland preservation across Canada reveals some interesting provincial approaches. Chapter 2 explored Quebec's two-pronged approach, putting emphasis on providing a broad provincial framework for farmland protection as well as ensuring sustainable agriculture in peri-urban areas. The broad provincial framework considers both the quality of farmland resources and the requirements for maintaining a viable agricultural sector. In peri-urban areas, the province considers developing and maintaining socially and economically productive agricultural systems to be of utmost importance. The flexible approach in these areas considers local characteristics, alternative forms of agricultural production, and proactive involvement of farmers and their families.

In Chapter 3, Ontario's attempt to find a balance between often conflicting views on property rights versus the public good was examined. Decisions made today will fundamentally affect options available to future generations: "The ability to produce food, to regulate the system of production to reflect the values of society, to maintain the important economic contributions of agriculture, and to retain the important role that farmers play in managing the countryside are dependent on retaining farmers and the lands essential to their livelihood."

British Columbia's Agricultural Land Reserve (ALR), born from passage of the Land Commission Act, has had a huge impact on the province's development over the past forty years. In Chapter 4, the successes and challenges that this bold land-use policy has had since its inception in 1973 were examined. The program to safeguard the province's scarce farmland resource has resulted in a reduction of the estimated annual loss of as much as 6,000 hectares of prime agricultural land to about 600 hectares since establishment of the ALR. In addition to successfully preserving agricultural land, the program has been able to convince the public of its importance. As Smith writes, "An opinion survey in 1997 found that over 80 percent of British Columbians considered it to be unacceptable to remove land from the ALR for urban uses, and a survey done in 2008 found that 95 percent of British Columbians supported the ALR and its farmland preservation policies."

Looking to the future, Chapters 6, 7, 8, 9, and 10 considered evolving issues and responses in several Canadian jurisdictions. Chapter 6 took a detailed look at Ontario's Smart Growth initiative, which will set the framework for land-use planning in Ontario for many years. Currently, the initiative is best seen as primarily an urban strategy that requires a better balance among land, environment, economy, and community to adequately meet the needs of rural communities. The current initiatives are powerful in protecting agricultural lands and natural environments, but there is a substantial weakness when it comes to economic development and community building. However, the initiatives can bring stability to the agricultural land market that will benefit primary producers in the long run. It remains to be seen if a balance between urban and rural can be found in the inherent costs associated with land-use protection policies as applied to rural areas.

In Chapter 7, we returned to British Columbia, where four decades of agricultural land protection have proven that urban sprawl is not a necessary component of livable urban communities. In fact, the ALR in British Columbia has revealed a sustainable, healthy agricultural sector and the creation of compact, livable communities. As McNaney and Lang note, "Jurisdictions across Canada can look west to see how a long-term vision has resulted in both a thriving agricultural industry and communities consistently rated as the most livable in the world."

Chapter 8 considered how land-use policies in Ontario have affected the creation of non-farm development on agricultural land. Considering data accumulated over many years, the argument for farmland preservation is bolstered by examining the cumulative impact of severances for residential lots throughout agricultural areas: "The provincially led planning approach to identifying and protecting prime agricultural areas at the landscape scale is an important progression toward addressing the gap between private property interests and public values." Although the land-use policies have had beneficial impacts on overall preservation, more communication is required on the rationales behind these policies from the province to the municipalities and from the municipalities to individual landowners to enhance effective implementation of these policies.

Farming in the peri-urban area presents challenges not experienced in more rural areas. Chapter 9 examined how agricultural entrepreneurs have been continually adapting to take advantage of the opportunities that also exist in these dynamic areas. Brunet points out that "the main objective of these strategies is the generation of sufficient income to remain viable and productive while improving links with non-farm rural residents and urban communities."

Chapter 10 explored the workings of the Ontario Farmland Trust. In existence since 2004, the OFT represents a partnership among a number of Ontario stakeholders. Although urbanization continues to put pressure for development on agricultural land, leadership through policy, authentic community engagement, and permanent land protection can result in a collaborative approach to farmland preservation. As Setzkorn writes, "The Ontario Farmland Trust vision is a future in which the best farmland in Ontario is valued and permanently protected through sound policy, partnerships, and proactive community engagement; in which diverse farming communities thrive; and in which the protection of farmland, agriculture, and local food production is recognized as the foundation of a sustainable rural economy in Ontario."

The final three chapters examined farmland preservation in other countries. Chapters 11 and 12 considered farmland preservation policies and planning for agriculture in the United States. Although that country has traditionally placed a strong emphasis on planning for development, many communities now recognize the need to plan for farmland preservation. Achieving a stable farmland base is of utmost importance as population growth of more than 80 million people, mostly in metropolitan areas, is expected by 2050. Local and state funding as well as a variety of farmland protection techniques will be required to achieve that stable farmland base and preserve the best land to maintain food security for the future.

As discussed in Chapter 13, Australia struggles with the fact that its urban areas are located on its most productive lands, while vast swaths of less productive land remain uninhabited. Farmland preservation, for the most part, has not had a high national profile, and protection has been left up to the six states in the country: "Importantly, protecting productive potential does not appear to have been effective as a goal in itself; the multifunctional role of such land in reinforcing the aesthetics of the metropolitan area remains significant. To this long-standing issue has been added the more recent clashes among farming, mining, and water for agriculture and the environment."

As we come back to the questions raised at the outset of this conclusion, we hope that this book will help readers to have a deeper understanding of why farmland preservation is so important both for this generation and for those in the future, how farmland preservation is occurring today, and which issues are evolving as we move forward. Uncertainty arising from climate change, unknown costs and availability of energy, technological advances, and shifting political winds requires us to be vigilant and adaptable when considering the best approaches to farmland preservation. Future generations will expect nothing less of us.

CONTRIBUTORS

Nicolas Brunet is an Assistant Professor of Geography and Planning at the University of Saskatchewan. His work explores policies and practices that support improved governance of social ecological systems in Canada's northern regions.

Christopher R. Bryant is an Adjunct Professor of Geography at Université de Montréal and School of Environmental Design and Rural Development at University of Guelph.

Trevor Budge is a Senior Lecturer and Coordinator of the Master of Arts in Community Development, Bachelor of Urban, Rural and Environmental Planning, and the Graduate Diploma in Rural and Regional Planning at La Trobe University, Bendigo, Australia. He also holds an Adjunct Professor position at RMIT University, Melbourne, in the planning program.

Andrew Butt is a Senior Lecturer in the Community Planning and Development Program at La Trobe University, Bendigo, Australia. His teaching and research has a specific focus on changing rural and farming landscapes, socio-economic processes of peri-urbanization, and consideration of the roles of planners in this process.

Wayne Caldwell is a Professor in Rural Planning at the University of Guelph. He is a Registered Professional Planner and is a passionate advocate for the betterment of rural communities. He has served as chair or president of a number of local, provincial, and national organizations.

Ghalia Chahine is a Regional Coordinator for the Système alimentaire montréalais (SAM) (Food Security Montreal) based in Concertation Montréal.

Arthur Churchyard has worked in the area of rural and agricultural planning for a number of years. He currently works with the Environmental and Land Use Policy Branch of the Ontario Ministry of Agriculture, Food, and Rural Affairs.

Tom Daniels is a Professor in the Department of City and Regional Planning at the University of Pennsylvania.

Gary Davidson has many years of experience in rural planning as a public and private planner. He is the past president of both the Canadian Institute of Planners and the Ontario Professional Planners Institute.

Claire Dodds is a Senior Planner with the County of Huron in Ontario. In this capacity she works with rural communities planning for the agriculture. She holds a Master of Science in Rural Planning from the University of Guelph.

Denis Granjon is a Professor of Geography at the Collège Lionel-Groulx in Sainte-Thérèse, Quebec.

Jim Hiley is happily retired following a thirty-year career supporting applied agricultural land research initiatives from the local to international scales.

Stew Hilts is Professor Emeritus, Department of Land Resource Science, at the University of Guelph and the former chair of the Ontario Farmland Trust.

Kelsey Lang is a Registered Professional Planner in Ontario and a former graduate of the University of Guelph's MSc in Rural Planning and Development. She is passionate about new urbanism and currently resides on Long Island, in New York state.

Claude Marois is an Adjunct Professor of Geography at Université de Montréal.

Kevin McNaney is Director, North East False Creek Project Office, at the City of Vancouver. He was formally with Smart Growth BC and holds degrees from McGill University and the University of British Columbia.

Kate Procter holds a Bachelor of Science in Agriculture and a Master's of Science in Planning. As well as farming, she has worked as a freelance journalist, editor, author, and a professional consultant, which involved helping communities strategically plan to form a common vision to bring about action.

Matt Setzkorn is the Manager of Land Programs and Policy at the Ontario Farmland Trust, and has been with the organization for over seven years. He also farms near Guelph, Ontario, raising beef cattle, growing fruit, and running a winery.

Barry E. Smith recently retired from his position as a policy planner with the Agricultural Land Commission in British Columbia.

Michael Troughton was a Professor in Geography at Western University in London, Ontario. He spent much of his career studying rural and agricultural land-use issues in Canada and beyond. Sadly, Michael passed away in 2007. His chapter has been updated to reflect evolving trends related to farmland in Canada.

Bob Wagner served as a Senior Associate for Farmland Protection Policy for American Farmland Trust, a U.S. non-profit organization dedicated to farmland protection and environmental stewardship. He has recently retired from this position.

Bronwynne Wilton holds a PhD in Rural Studies and a Master's of Science in Rural Planning/Landscape Architecture. Bronwynne is currently a Consulting Project Lead with Synthesis Agri-Food Network in Guelph, Ontario.